Biodiesel Production with Green Technologies

Aminul Islam • Pogaku Ravindra

Biodiesel Production with Green Technologies

 Springer

Aminul Islam
Faculty of Engineering
Universiti Malaysia Sabah
Kotakinabalu, Malaysia

Energy Research Unit
University Malaysia Sabah
Kotakinabalu, Malaysia

Pogaku Ravindra
Faculty of Engineering
Universiti Malaysia Sabah
Kotakinabalu, Malaysia

Energy Research Unit
University Malaysia Sabah
Kotakinabalu, Malaysia

ISBN 978-3-319-83256-2   ISBN 978-3-319-45273-9   (eBook)
DOI 10.1007/978-3-319-45273-9

Printed on acid-free paper

This Springer imprint is published by Springer Nature
The registered company is Springer International Publishing AG
The registered company address is: Gewerbestrasse 11, 6330 Cham, Switzerland

*This book is dedicated to my beloved
parents, Md. Korban Ali and Mst Mahamuda
Begum, with love.
Actually, my parents don't read book,
so if someone doesn't tell them about this,
they'll never know.*

Aminul Islam

*Dedicated to two noble souls of my life
my divine parents
"Babha Anjaiah Pogaku
and Amma Shantha bai"*

Pogaku Ravindra

Abstract

Renewable energy resources appear to be one of the most efficient and effective solutions. Thus, it is essential to replace gradually the current nonrenewable resources with renewable ones to meet future demand of energy. Many of the changes are having a profound influence on renewable energy, from policies explicitly designed to promote renewable energy. Among the many renewable fuels currently available around the world, biodiesel offers an immediate impact in our energy. Biodiesel is a renewable, biodegradable, and nontoxic fuel. Biodiesel production using various types of heterogeneous metal oxide catalysts has been studied in the past. However, most of these catalysts have been prepared in the form of powders with sizes ranging from nano- to micrometer. The small particle size may offer high catalytic activity, but it gives rise to several problems such as high pressure drops, poor mass/heat transfer, poor contact efficiency, and difficulties in handling and separation. From the practical point of view, handling of small particles could be difficult due to the formation of pulverulent materials. Therefore, attention has been paid to the selection of green catalytic process in this book where the catalyst could be highly selective towards the formation of a desired product. Furthermore, the catalyst could be easily handled, recovered from the reaction medium and subsequently reused. The main focus is on the description of the state of the art on catalytic processes that are expected to play a decisive role toward the "green" production of biodiesel. Our effort in this area are directed toward understanding the mechanisms involved in the synthesis and structure formation of catalyst in order to get high yield of biodiesel production. At the same time, this book addresses the question of how catalytic material should be distributed inside a porous support to obtain optimal performance. This understanding can be used to control the microstructure of the catalyst, and hence the properties of catalyst. The effects of physicochemical and operating parameters are analyzed to gain insight into the underlying phenomena governing the performance of optimally designed catalysts. A balance description of theory and experiment and stress problems of commercial importance have also been emphasized in this book.

Contents

Chapter 1
Introduction

The enormous worldwide use of diesel fuel and the rapid depletion of crude oil reserves have prompted keen interest and exhaustive research into suitable alternative fuel. Currently, attention is focused on human and environmental safety, in relation to the release of hydrocarbons into the environment. Petroleum derivatives contain benzene, toluene, ethylbenzene, and xylene isomers, the major components of fossil fuel, which are hazardous substances subject to regulations in many parts of the world (Serrano, Gallego, & Gonzalez, 2006). As a consequence, the demand of green energy is increasingly gaining international attention. When green energy is used, the primary objective is to reduce air pollution, and minimize or eradicate completely any impacts to the environment (Burgess, 1990). Among many possible sources, apparently, biodiesel is a viable alternative energy to conventional diesel fuel, which is of environmental concern and is under legislative pressure to be replaced by biodegradable substitutes.

The most common way to produce biodiesel is by transesterification which refers to a catalyzed chemical reaction involving vegetable oil and an alcohol to yield fatty acid alkyl esters (biodiesel) and glycerol (Freedman & Pryde, 1984; Lotero et al., 2006), as shown in Fig. 1.1. Triglycerides, as the main component of vegetable oil, consist of three long chain fatty acids esterified to a glycerol structure. When triglycerides react with an alcohol (e.g., methanol), the three fatty acid chains are released from the glycerol skeleton and combine with the methanol to yield fatty acid methyl esters (FAME). Glycerol is produced as a by-product. The transesterification reaction can be carried out using homogeneous, heterogeneous, or enzymatic catalysts (Dossin, Reyniers, & Marin, 2006; Jegannathan, Abang, Poncelet, Chan, & Ravindra, 2008; Lopez, Goodwin, Bruce, & Lotero, 2005).

Homogeneous catalysts (e.g., NaOH and KOH) are usually employed commercially for the preparation of biodiesel. Some of the shortcomings include the formation of an unwanted soap by-product in the presence of water and free fatty acids (FFA). From a process perspective, homogeneous catalysts, however, are corrosive,

© Springer International Publishing Switzerland 2017
A. Islam, P. Ravindra, *Biodiesel Production with Green Technologies*,
DOI 10.1007/978-3-319-45273-9_1

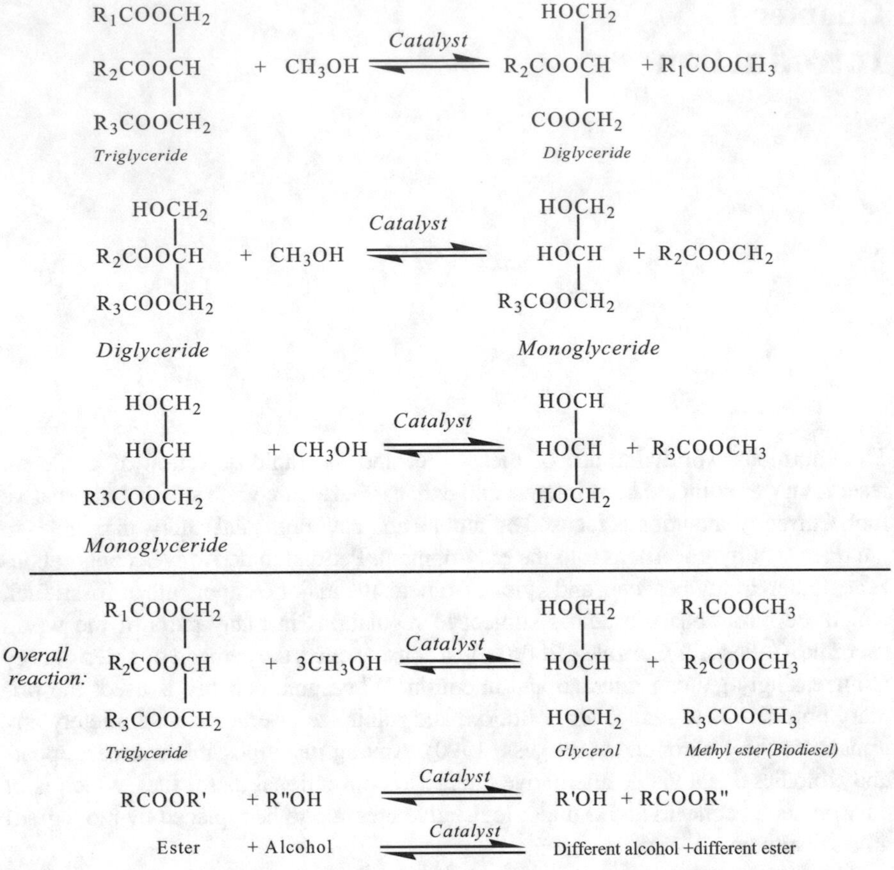

$$
\begin{array}{ll}
\text{R}_1\text{COOCH}_2 & \text{HOCH}_2 \\
| & | \\
\text{R}_2\text{COOCH} \quad + \text{CH}_3\text{OH} \xrightleftharpoons{\text{Catalyst}} \text{R}_2\text{COOCH} \quad + \text{R}_1\text{COOCH}_3 \\
| & | \\
\text{R}_3\text{COOCH}_2 & \text{COOCH}_2 \\
\textit{Triglyceride} & \textit{Diglyceride}
\end{array}
$$

$$
\begin{array}{ll}
\text{HOCH}_2 & \text{HOCH}_2 \\
| & | \\
\text{R}_2\text{COOCH} \quad + \text{CH}_3\text{OH} \xrightleftharpoons{\text{Catalyst}} \text{HOCH} \quad + \text{R}_2\text{COOCH}_2 \\
| & | \\
\text{R}_3\text{COOCH}_2 & \text{R}_3\text{COOCH}_2 \\
\textit{Diglyceride} & \textit{Monoglyceride}
\end{array}
$$

$$
\begin{array}{ll}
\text{HOCH}_2 & \text{HOCH} \\
| & | \\
\text{HOCH} \quad + \text{CH}_3\text{OH} \xrightleftharpoons{\text{Catalyst}} \text{HOCH} \quad + \text{R}_3\text{COOCH}_3 \\
| & | \\
\text{R3COOCH}_2 & \text{HOCH}_2 \\
\textit{Monoglyceride} &
\end{array}
$$

$$
\begin{array}{lll}
& \text{R}_1\text{COOCH}_2 & \text{HOCH}_2 \quad \text{R}_1\text{COOCH}_3 \\
\textit{Overall} & | & | \\
\textit{reaction:} & \text{R}_2\text{COOCH} \quad + 3\text{CH}_3\text{OH} \xrightleftharpoons{\text{Catalyst}} \text{HOCH} \quad + \text{R}_2\text{COOCH}_3 \\
& | & | \\
& \text{R}_3\text{COOCH}_2 & \text{HOCH}_2 \quad \text{R}_3\text{COOCH}_3 \\
& \textit{Triglyceride} & \textit{Glycerol} \quad \textit{Methyl ester(Biodiesel)}
\end{array}
$$

$$
\text{RCOOR}' + \text{R''OH} \xrightleftharpoons{\text{Catalyst}} \text{R'OH} + \text{RCOOR''}
$$

$$
\text{Ester} + \text{Alcohol} \xrightleftharpoons{\text{Catalyst}} \text{Different alcohol +different ester}
$$

Fig. 1.1 Overall transesterification reaction. Source: Lotero et al. (2006)

can be used only once, and require energy intensive separation operations that lead to waste formation and environmental pollution (Meher, Kulkarni, Dalai, & Naik, 2006). On the other hand, the enzymatic catalysts offer considerable advantages when compared with both acidic and alkaline homogeneous catalysts. Those have, for instance, less sensitivity to water, increased catalyst recovery, and better efficiency in the biodiesel separation steps (Akoh et al., 2007). However, the drawbacks for the use of enzymes are: (a) low reaction rate (Zhang, Dube, McLean, & Kates, 2003); (b) their cost (Meher, Kulkarni, et al., 2006) for industrial-scale use 1,000 US$ per kg compared to 0.62 US$ (Haas, McAloon, Yee, & Foglia, 2006) for sodium hydroxide.

On the contrary, the use of heterogeneous catalytic systems in the transesterification of triglyceride implies the elimination of several steps of washing of biodiesel, ensuring thereby higher efficiency and profitability of the process as well as lowering the production cost, as summarized by Kawashim et al. (2008). However, the development of a stable catalyst with suitable particle size that can be recycled and reused to simplify the product separation and purification steps remains inadequately addressed. Therefore, a number of researches have been directed towards the development of heterogeneous catalyst for transesterification of triglycerides over the past decade.

Many heterogeneous catalysts for the transesterification of oils have been developed. For example, the transesterification reaction of soybean oil with KI/Al_2O_3 has been studied; conversion in excess of 90 % was achieved at a temperature of 60 °C (Suppes, Dasari, Doskocil, Mankidy, & Goff, 2004). It has also been reported that the conversion to methyl ester reaches 87 % with the potassium-loaded alumina catalyst, when a mixture with a molar ratio of methanol to oil of 15:1 is refluxed for a reaction time 7 h (Xie, Peng, & Chen, 2006b). Besides these, a great variety of γ-Al_2O_3 supported catalysts including $Mg(NO_3)_2/Al_2O_3$ (Benjapornkulaphong, Ngamcharussrivichai, & Bunyakiat, 2009), $Na/NaOH/\gamma$-Al_2O_3 (Kim et al., 2004), $NaOH/Al_2O_3$ (Arzamendi et al., 2007), KNO_3/Al_2O_3 (Vyas, Subrahmanyam, & Patal, 2009; Xie, Peng, & Chen, 2006a), $K/KOH/\gamma$-Al_2O_3 (Ma, Li, Wang, Wang, & Tian, 2008) have been investigated under various reaction conditions and with a variable degree of success. However, information on the catalyst reusability in transesterification reaction is still rather limited.

More recently, the utilization of Al_2O_3 supported KF catalyst having a particle size of nanometer order for biodiesel production has been demonstrated by Boz, Degirmenbasi, and Kalyon (2009) and reported that the catalyst can be effectively used in the production of biodiesel from vegetable oil. Wang et al. reported that nano-MgO was used in the transesterification of soybean oil as a catalyst in supercritical and subcritical methanol. Further, Wen, Wang, Lu, Hu, and Han (2010) reported that high yield of biodiesel (97 %) can be obtained in the transesterification of Chinese tallow seed oil at 65 °C after 2.5 h of reaction using KF/CaO nanocatalyst. Besides, catalysts with **nano**-sized such as CaO.ZnO (Ngamcharussrivichai, Tuturat, & Bunyakiat, 2008), K_2O/Al_2O_3 (Han & Guan, 2009) have been demonstrated to be efficient heterogeneous catalyst for transesterification.

The significant step forward has come more recently with the discovery of the mesoporous M41S family in 1992 (Kresge, Leonowicz, Roth, Vartuli, & Beck, 1992), which offer many opportunities over microporous materials by being more accessible to reactants. Recently, the use of mesoporous catalysts such as zeolites (Sasidharan & Kumar, 2004), SBA-15 (Ignat et al., 2010; Pariente, Dıaz, Mohino, & Sastre, 2003; Shah, Ramaswamy, Lazarc, & Ramaswamy, 2004), Ti-HMS (Shuwen et al., 2007), MCM-41 (Liu et al., 2008d; Pariente et al., 2003; Jr. et al., 2009), Al-MCM-41 (Jr. et al., 2009; Rashtizadeh, Farzaneh, & Ghandi, 2010) has been reported for the transesterification of vegetable oil. Some of these research findings had proven the potential application of catalyst for transesterification reactions; however, from a practical point of view, handling of small particles in large quantities is difficult and limits the possibilities to recover for reuse. Thus, the simplicity of catalyst reusability along with their handling requirements is a crucial prerequisite for designing a supported catalyst.

A variety of supported porous catalysts for transesterification of vegetable oils have been investigated in laboratory scale, including K/SBA-15 (Abdullah, Razali, & Lee, 2009), KNO_3/MCM-48 (Sun et al., 2009), KI/mesoporous silica (Samart, Sreetongkittikul, & Sookman, 2009), CaO/mesoporous silica (Albuquerque et al.,

2008; Xin, Zhen, Hua, & Yong, 2009). Other inorganic oxides supported catalyst applied for transesterification of vegetable oil are NaN_3/mesoporous γ-Al_2O_3 (Bota, Houthoofd, Grobet, & Jacobs, 2010), KF, CsF, LiF supported on mesoporous alumina (Verziu et al., 2009), MgAl hydrotalcites (Tantirungrotechai et al., 2010). However, these catalysts preparation approaches to be rather complicated, which limits their potentialities in large scale application. Concerning the catalyst properties, it has been reported by Perego and Villa (1997) that the catalyst should have a high surface area and desirable mechanical strength to increase its stability under reaction condition and hence the catalyst life. In catalysis, gamma alumina is preferred because it can provide high surface area and thermal stability, which are suitable characteristics for the catalytic performance (Chuah, Jaenicke, & Xu, 2000).

Although a variety of solid catalysts have been studied with varying degrees of success, these catalysts have been prepared in the form of powders with diameter ranging from nanometer to micrometer. From the practical point of view, handling of small particles in large quantities could be difficult due to the formation of pulverulent materials. There are also possible health risks caused by inhalation of small particles. Utilization of powders in conventional catalytic reactions renders their recovery and purification difficult, and ultracentrifugation is needed for the subsequent separation. In particular, the use of powder catalysts gives rise to a number of problems on an industrial scale, including high pressure drops (Centi & Perathoner, 2003a, 2003b); high diffusion resistance (Williams, 2001); poor mass/heat transfer, and is not amenable to scale up (Meille, 2006). The particles in the form of powder could be plugged in the reactor or formed the aggregated cluster in the reaction medium which might reduce the mass transfer/heat transfer from the reactant to the active site of the catalyst (Meille, 2006). In fact, the catalytic activity might be reduced. In addition, diffusion is the process, in which the ions moves from the solution to the active surface of the catalyst particle. As reported by Williams, (2001) that the smaller the particle size, the greater will be the resistance against which an ion must flow inside the active surface of particle. In addition, the active phase of the catalyst may not be uniformly distributed on the support but rather form localized aggregates leading to low contact of active surface in the catalyst (Wang et al., 2008). Thus, the efficiency of the catalyst and its feasibility at industrial scale might be reduced. Therefore, the design of a catalyst form at a macroscale (millimeters in diameter) is indispensable to avoid the problems in relation to the traditional catalysts.

Despite the many potential advantages using biodiesel, there are two major obstacles relevant to the cost of current biodiesel production; (a) the limited availability and volume of oil feedstocks needed to supply the current demand for diesel fuel, and (b) the processing cost. Furthermore, there has been no subsequent work to improve the biodiesel yield or to demonstrate the reusability of using macroscopic catalyst. The current biodiesel plants uses homogeneous alkali catalysts operating in a batch-type process, followed by an additional effort to remove the liquid catalysts and saponified products (Mekhilef, Siga, & Saidur, 2011; MPOB, 2007). In addition, the process would eliminate the handling issues of hazardous liquid base catalysts. Furthermore, it would minimize the separation steps, and bring more benefits associated with it; for instance, it is easy to be recovered, and can potentially be reused. Eventually, the production costs can be substantially reduced.

The main reason to consider heterogeneous catalysis for industrial processes is the ease of catalyst separation after the reaction. Separation processes represent more than half of the total investment in equipment for the chemical and fuel industries (King, 2007). It is reasonable to state that the separation costs are a decisive factor in the final analysis of a new process. Therefore, the ease of separation of solid catalysts can be a crucial advantage. However, most of the catalysts employed for transesterification process are in the form of a powder ranging from micrometer to nanometer in diameter. Despite the high percentage of fatty acid methyl ester (FAME) (>90 %) achieved by powdered catalysts, many catalytic systems have not been commercialized because of the difficulties encountered when trying to separate such catalysts from the reaction media (Cole-Hamilton, 2003). In addition, from the practical point of view, handling of such small particles in large quantities could be difficult due to the formation of dust. Furthermore, the smaller the emitted particle, the more harmful it is to the human body because particles under 100 nm (ultrafine particles) in diameter have a higher surface area per unit mass of particles; therefore, the smaller particles can easily infiltrate into the respiratory organs (Donaldson, Li, & MacNee, 1998). Thus, the utilization of millimetric spherical catalyst support and their application in transesterification reactions will be a practical alternative to catalyst in powder form in view of following advantages; (1) easy of separation of the catalyst by simple filtration, (2) the catalyst is easy to handle and reusable for several cycles, and (3) the present method is simple over the existing procedures. Therefore, the problems associated with the conventional catalyst would preferentially be replaced by using millimetric spherical supported catalyst which will be environmentally friendly and simplify the existing processes of biodiesel production.

Chapter 2
Literature Review

2.1 Introduction

The expansive use of diesel fuel worldwide and the rapid depletion of crude oil reserves have prompted keen interest and exhaustive research into suitable alternative fuels. Biomass sources, particularly vegetal oils, have attracted much attention in recent years because of their wide availability and ease of renewal. Alternative fuels for diesel engines are becoming increasingly important because of the diminishing petroleum reserves. In addition, environmental pollution, through the emission of carbon monoxide, sulfur dioxide, hydrocarbons, and hazardous particulates, and the threat of climatic change associated with green house effect are the most serious problems across the world (Amigun, Sigamoney, & Blottnitz, 2008). Therefore, the demand for clean, alternative fuel has been growing rapidly. Among the many possible sources, biodiesel attracts the most attention as a promising substitute for conventional diesel fuel. Biodiesel refers to the lower alkyl esters of long chain fatty acids, which are synthesized either by transesterification with lower alcohols or by esterification of fatty acids in the presence or absence of a catalyst (Freedman & Pryde, 1984).

This chapter reviews a brief overview of the different type of catalyst used in the process of transesterification of oils for the production of biodiesel with special reference to the heterogeneous catalyst. The heterogeneous catalyst preparation and impregnation methods are included in this section. The solid catalysts used, together with the different kinds of feedstock, process variables, reaction conditions, and reusability of the catalysts, are also discussed. The summary is indicated in the following flowchart (Fig. 2.1).

© Springer International Publishing Switzerland 2017
A. Islam, P. Ravindra, *Biodiesel Production with Green Technologies*,
DOI 10.1007/978-3-319-45273-9_2

Fig. 2.1 Summary of
literature review

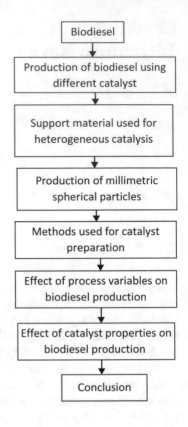

2.2 Biodiesel

Biodiesel (fatty acid alkyl esters) is an alternative fuel similar to conventional or fossil diesel. Biodiesel is made from renewable resources like vegetable oil, animal oil/fats, tallow, and waste cooking oil. Rudolf Diesel developed the first diesel engine which was run with vegetable oil in 1900. After eight decades, the awareness about environment rose among the people to search for an alternative fuel that could burn with less pollution. It was reported by Gerpen and Knothe (2005) that a decreased dependence on foreign sources of fuel will enhance national security, interest in the use of biodiesel as an alternative fuel has accelerated.

The advantages of biodiesel as diesel fuel are:

(a) Biodegradablity (Mittelbach & Remschmidt, 2004);
(b) Lower sulfur and aromatic content (Ma & Hanna, 1999);
(c) Ready availability, renewability, higher combustion efficiency (Ma & Hanna, 1999);
(d) High flash combustion temperature (Knothe, Krahl, & Van Gerpen, 2005);
(e) Neutral with regard to carbon dioxide emissions (Knothe, Sharp, & Ryan, 2006).

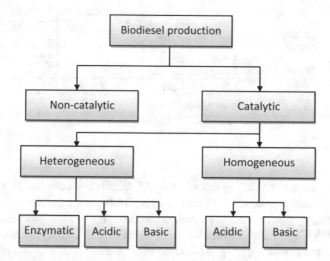

Fig. 2.2 Classification of catalyst used for biodiesel production

Biodiesel is usually produced by converting the triglyceride oils with an alcohol to esters with a process known as transesterification shown in Fig. 1.1. The transesterification process can be classified into two broad catagories; namely catalytic and non catalytic process (Fig. 2.2). The scope of this thesis is limited to heterogeneous catalytic process; however, an overview of the transesterification process and their impact on biodiesel production are discussed in the following sections.

2.2.1   Non-catalyzed Biodiesel Production

In a more recent development, non-catalytic supercritical transesterification with co-solvent provides a new way of producing biodiesel fuel from bio-based oils (triglycerides) (Gui, Lee, & Bhatia, 2009).

It was reported that biodiesel can be produced at a relatively fast rate without the presence of catalyst by heating up to supercritical stage of methanol, ethanol, propanol and butanol (Demirbas, 2006). The critical temperatures and critical pressures of the various alcohols are shown in Table 2.1.

Many methods have been proposed for biodiesel production in supercritical (SC) technology including- SC methanol (He, Sun, Wang, & Zhu, 2007a), SC ethanol (Gui et al., 2009), SC methanol with CO_2 as a co-solvent (Kasim, Tsai, Gunawan, & Ju, 2009), and SC carbon dioxide with enzyme (Madras, Kolluru, & Kumar, 2004). He, Sun, et al. (2007a)) subjected soybean oil to transesterification process in the absence of catalyst with the supercritical methanol and found that the maximum biodiesel yield can be obtained at 310 °C. However, it has been reported that the side reactions of unsaturated fatty acid methyl esters (FAME) at reaction temperature above 300 °C leads to much loss of material (He, Wang, & Zhu, 2007b). As reported by He, Wang,

Table 2.1 Critical
temperatures and critical
pressures of various alcohols

Alcohol	Critical temperature (°C)	Critical pressure (bar)
Methanol	239.2	81
Ethanol	243.2	64
1-Propanol	264.2	51
1-Butanol	287.2	49

Source: Demirbas, 2006

et al. (2007b) that gradual heating procedure could effectively reduce the loss of materials caused by the side reactions of unsaturated FAME, but the high temperature is still very much energy intensive for biodiesel production. To solve this problem, the use of methanol with CO_2 as a co-solvent has been suggested, since they are able to reduce supercritical temperature during production of biodiesel (Kasim et al., 2009). Since supercritical CO_2 is a good solvent for vegetable oil, it allows the reaction mixture to form a single phase which will accelerate the reaction rate at relatively lower temperature (Pinnarat & Savage, 2008).

The influence of process parameters has been investigated for non-catalytic transesterification to optimize the biodiesel content (Demirbas, 2007). Hawash, Kamal, Zaher, Kenawi, and Diwani (2009) also found that the conversion of Jatropa oil almost tripled when the temperature increased from 239 to 340 °C. Similar results have been reported by Ilham and Saka (2009). After 12 min of reaction at 350 °C and pressure 200 bar, rapeseed oil treated with supercritical dimethyl carbonate reached 94 % yield of fatty acid methyl ester. Likewise, the conversion was 96 % in case of coconut oil and palm kernel oil over the same temperature range with molar ratio of methanol to oil 4 and the reaction time of 10 min (Bunyakiat, Makmee, Sawangkeaw, & Ngamprasertsith, 2006). Other results consistent with these findings are from Silva et al. (2007). However, the conversion below the solvent critical temperature is very low (Silva et al., 2007). It should be noted that the critical temperatures of methanol and ethanol are 240 and 243 °C, respectively, and, therefore, the conditions at 200 °C represent a subcritical state of the medium, as classified by Madras et al. (2004).

It was reported by Yin, Xiao, and Song (2008) that the yield drastically increased when the temperature was changed from subcritical to supercritical state for transesterification with methanol or ethanol. This could be due to the absence of mass transfer interphase under these conditions to limit the reaction rate. However, the temperature beyond the supercritical, the FAME yield could be dropped, as reported by Gui et al. (2009). According to Gui et al. (2009), at temperatures above supercritical temperature, the unsaturated fatty acids such as oleic acid and linoleic acid tends to decompose via isomerization of the double bond functional group from cis-type carbon bonding (C=C) into trans-type carbon bonding (C=C), which is naturally unstable fatty acids. Therefore, it was highly possible to drop FAME yield with increasing temperature especially at conditions beyond the supercritical point (Gui et al., 2009).

The influence of different alcohols on supercritical transesterification has also been investigated. For instance, in methanol, the conversion increased from 78 to

Table 2.2 Summarization of non-catalyzed biodiesel production

M/O	Oil	RT (°C)	MFP	FY (%)	P (bars)	SF	References
40:1	Sunflower	350	1.45	96	200	SCM	Madras et al., 2004
41:1	Rapeseed	350	16	95	450	SCM	Saka & Dadan, 2001
40:1	Soybean	310	1.87	77	250	SCE	He, Wang, et al., 2007b
41:1	Cottonseed	230	7.53	98	-	SCM	Demirbas, 2008
41:1	Cottonseed	230	5.76	75	-	SCE	Demirbas, 2008
42:1	Soybean	280	1.8	90	350	SCM	He, Sun, et al., 2007a
271:1	Rice bran oil	300	-	51.28	300	SCM	Kasim et al., 2009
271:1	DDRBO	300	-	94.84	300	SCM	Kasim et al., 2009
M/O	Oil	RT (°C)	MFP	FY (%)	P (bars)	SF	References
41:1	Linseed	287	1.992	99.6	-	SCM	Demirbas, 2009
33:1	Palm	349	1.584	79.2	-	SCE	Gui et al., 2009
42:1	Rapeseed	350	4.7	94	200	SDC	Ilham & Saka, 2009
43:1	Jatropha	320	16.66	100	84	SCM	Hawash et al., 2009
41:1	Linseed	250	7.538	98	-	SCM	Demirbas, 2009
41:1	Sunflower	252	2.204	97	81	SCM	Demirbas, 2007

Maximum FAME productivity (MFP)=FAME (g)/Oil (g). t (h), FAME yield (FY)=FAME (g)/Oil (g), reaction temperature (RT), pressure (P), supercritical fluid (SF), supercritical dimethyl carbonate (SDC), supercritical ethanol (SCE), supercritical methanol (SCM), dewaxed-degummed rice bran oil (DDRBO)

96% with the increase in temperature (Saka & Dadan, 2001). A similar trend was observed for conversions in ethanol but the conversions were higher. Higher conversions in ethanol may be attributed to the solubility of the oil in the system (Kusdiana & Saka, 2001). Because the solubility parameter of ethanol is lower than that of methanol and is closer to the solubility parameter of the oil, the conversions are higher in ethanol compared to the conversions obtained in methanol (Madras et al., 2004). A number of studies have been focused on the costs of investment under the supercritical conditions. As reviewed by Pinnarat and Savage (2008), transesterification is much simpler and more environmentally friendly because of the absence of catalyst, the purification of products after transesterification. However, reaction requires temperatures of 350–400 °C, pressures of 200–450 bar, and methanol to methanol to oil ratio of 40–271, as summarized in Table 2.2, which are not viable in practice in industry due to drive up some aspects of the processing cost.

2.2.2 Homogeneous Base-Catalyzed Processes

The transesterification process is catalyzed by alkaline metal alkoxides (Schwab, Bagby, & Freedman, 1987) and hydroxides (Dmytryshyn, Dalai, Chaudhari, Mishra, & Reaney, 2004) as well as sodium or potassium carbonates (Korytkowska,

Barszczewska-Rybarek, & Gibas, 2001). Leung and Guo tested NaOH, KOH, and CH₃ONa as catalysts for biodiesel synthesis. Alkaline metal alkoxides (as CH₃ONa for the methanolysis) are the most active catalysts. They give yields greater 98 % in a relatively short reaction time of 30 min even at low molar concentrations of about 0.5 mol%, but their requirement of the absence of water makes them inappropriate for typical industrial processes in which water cannot be avoided completely (Freedman & Pryde, 1984).

Alkaline metal hydroxides (e.g., KOH and NaOH) are cheaper than metal alkoxides, but less active. Nevertheless, they are a good alternative since they can give the same high conversions of vegetable oils just by increasing the catalyst concentration to 1 or 2 mol%. However, even if water-free alcohol–oil mixture is used, some water is produced in the system by the reaction of the hydroxide and the alcohol. The presence of water gives rise to hydrolysis of some of the produced ester (Fig. 2.3), with consequent soap formation (Freedman & Pryde, 1984).

In 2003, metal complexes of the type M(3-hydroxy-2-methyl-4-pyrone)₂(H₂O₂), where M = Tin(Sn), Zinc (Zn), Lead (Pb), and Mercury (Hg), were used for soybean oil methanolysis under homogeneous conditions (Abreu et al., 2003). The Sn complex at a molar ratio of 400:100:1 methanol–oil–catalyst gave 90 % conversion in 3 h, while the Zn complex gave only 40 % conversion under the same conditions. This undesirable saponification reaction reduces the yield of ester and makes the recovery of the glycerol considerably more difficult due to the formation of emulsions, increase in viscosity, and greatly increased product separation cost. Potassium carbonate, used in a concentration of 2 or 3 mol% gives high yields of fatty acid alkyl esters and reduces the soap formation (Filip, Zajic, & Smidrkal, 1992). This can be explained by the formation of bicarbonate instead of water, which does not hydrolyze the esters.

Base-catalyzed transesterification is the most commonly used technique as it is the most economical process since it requires only low temperatures and pressures, and produces over 98 % conversion yield (provided the starting oil is low in moisture

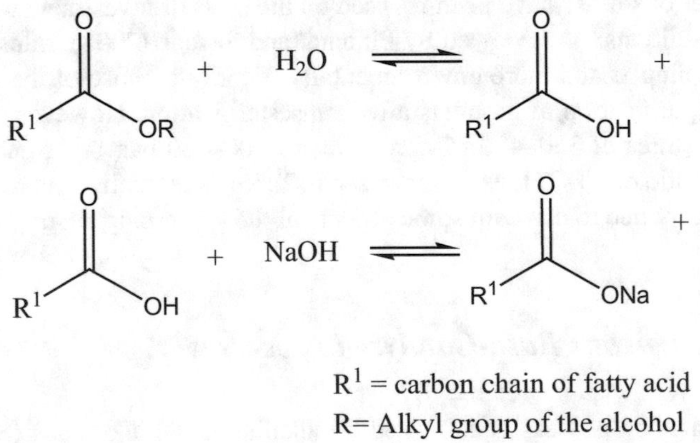

R^1 = carbon chain of fatty acid
R = Alkyl group of the alcohol

Fig. 2.3 Hydrolysis of ester and formation of soap by the presence of water

and FFA) (Singh, He, Thompson, & Van Gerpen, 2006). Base-catalyzed transesterification, however, has some limitations among which are that it is sensitive to FFA content of the feedstock oils. The presence of water and high amount of free acid gives rise to saponification of oil and therefore, incomplete reaction during alkaline transesterification process with subsequent formation of emulsion and difficulty in separation of glycerol (Leung & Guo, 2006). Other drawback of the base-catalyzed transesterification is that the process is energy intensive, alkaline catalyst has to be removed from the product and alkaline waste water requires treatment (Meher, Sagar, & Naik, 2006).

2.2.3 Homogeneous Acid-Catalyzed Processes

The transesterification process is catalyzed by sulfuric (Goff, Bauer, Lopes, Sutterlin, & Suppes, 2004), hydrochloric (Lee, Park, Lim, Han, & Lee, 2000), and organic sulfonic acids (Stern & Hillion, 1990). Preferably, sulfonic and sulfuric acids are mostly used. These catalysts give very high yields in alkyl esters, but the reactions are slow, requiring typically, temperatures above 100 °C and from 3–48 h to reach complete conversion (Zheng, Kates, Bube, & McLean, 2006).

Freedman and Pryde (1984) showed that the methanolysis of soybean oil, in the presence of 1 mol% of H_2SO_4, with an alcohol–oil molar ratio of 30:1 at 65 °C, takes 50 h to reach complete conversion of the vegetable oil (>99 %), while the butanolysis (at 117 °C) and ethanolysis (at 78 °C) using the same quantities of catalyst and alcohol take 3 h and 18 h, respectively.

Reaction rates in acid-catalyzed processes may be increased by the use of larger amounts of catalyst. Typically, catalyst concentrations in the reaction mixture have ranged between 1 and 5 wt% in most studies using sulfuric acid (Freedman, Butterfield, & Pryde, 1986). Canakci and Gerpen (1999) used different amounts of sulfuric acid (1, 3, and 5 wt%) in the transesterification of grease with methanol. In these studies, a rate enhancement was observed with the increased amounts of catalyst and ester yield went from 72.7 to 95.0 % as the catalyst concentration was increased from 1 to 5 wt%. The dependence of reaction rate on catalyst concentration has been further verified by the same authors and other groups (Crabba, Nolasco-Hipolito, Kobayashi, Sonomoto, & Ishizaki, 2001). A further complication of working with high acid catalyst concentration becomes apparent during the catalyst neutralization process, which precedes product separation.

Also, hydrochloric, organic, sulfonic, formic, acetic, and nitric acids have been investigated by other authors (Lotero et al., 2005). The acid-catalyzed process is thought to be more suitable for the production of biodiesel from low feedstocks (used frying oil, waste animal fat), mainly because of the fact that these feedstocks contain greater amounts of free fatty acids (FFAs) (Siakpas, Karagiannidis, & Theodoseli, 2006). The greater tolerance of an acid catalyst to the FFA content compared to an alkaline catalyst was confirmed in a report by Canakci and Van Gerpen (1999). They also showed that acid catalyzed reactions are more susceptible to water content of the feedstock than the base-catalyzed process. Also, there is evidence that large quantities of acid catalyst in biodiesel production may lead to ether formation by alcohol dehydration (Keyes, 1932) and the consequent high use of calcium oxide in the acid neutralization after production with its attendant high production cost and waste generation.

It has been suggested that acid-catalyzed transesterification achieves greater and faster conversions at high alcohol concentrations (Lotero et al., 2005). However, ester yields do not proportionally increase with molar ratio. For instance, for soybean methanolysis using sulfuric acid, ester formation sharply improved from 77 % using a methanol-to-oil ratio of 3.3:1–87.8 % with a ratio of 6:1. Higher molar ratios showed only moderate improvement until reaching a maximum value at a 30:1 ratio (98.4 %) (Lotero et al., 2006). Freedman and Pryde (1984) investigated the transesterification of soybean oil with methanol using 1 wt% concentrated sulfuric acid (based on oil). They found that at 65 °C and a molar ratio of 30:1 methanol to oil, it took 69 h to obtain more than 90 % oil conversion to methyl esters.

Acid-catalyzed processes are insensitive to fatty acid ester (FFA) and are better than the alkaline catalysts for vegetable oils with higher FFA. Thus, a great-advantage with acid catalysts is that they can directly produce biodiesel from low-cost lipid feedstock generally associated with high FFA concentrations, including waste frying oils. Despite its insensitivity to free fatty acids in the feedstock, acid-catalyzed transesterification has been largely ignored mainly because of its relatively slower reaction rate (Zhang, Dube, McLean, & Kates, 2003). For acid catalyzed conversion of waste vegetable oil with high free fatty acid content, higher alcohol to oil ratio is required compared to basic catalyzed operation for better yield of biodiesel. Other disadvantages with this process are acidic effluent, no reusable catalyst and high cost of equipment (Wang, Ou, Liu, Xue, & Tang, 2006).

All these factors in addition to the serious environmental and corrosion-related problems limit their use. Both the base-catalyzed and the acid-catalyzed transesterification processes have their advantages and disadvantages as previously mentioned. Hence, to avoid the problems associated with the use of these catalysts separately, current research has been directed toward the development of heterogeneous catalysts for biodiesel production.

2.2.4   Enzyme-Catalyzed Transesterification

There is a current interest in using enzymatic catalysis to commercially convert vegetable oils and fats to FAME as biodiesel fuel, since it is more efficient, highly selective, involves less energy consumption (reactions can be carried out in mild conditions) and produces less side products or waste (environmentally favorable) (Akoh et al., 2007). The transesterification process is catalyzed by lipases such as *Candida antarctica* (Royon, Daz, Ellenrieder, & Locatelli, 2007), *Candida rugosa* (Linko et al.,

1998), Pseudomonas cepacia (Ghanem, 2003), immobilized lipase (Bernardes, Bevilaqua, Leal, Freire, & Langone, 2007), and Pseudomonas sp. (Lai, Ghazali, & Chong, 1999).

The enzymatic alcoholysis of soybean oil with methanol and ethanol was investigated using a commercial, immobilized lipase (Bernardes et al., 2007). In that study, the best conditions were obtained in a solvent-free system with ethanol–oil molar ratio of 3.0, temperature of 50 °C, and enzyme concentration of 7.0 % (w/ w). They obtained yield 60 % after 1 h of reaction. In another study, Shah and Gupta (2007) obtained a high yield (98 %) by using P. cepacia lipase immobilized on celite at 50 °C in the presence of 4–5 % (w/w) water in 8 h. A more recent study by Maceiras, Vega, Costa, Ramos, and Marquez (2009) was also conducted to investigate the enzymatic conversion of waste cooking oils into biodiesel using immobilized lipase Novozym 435 as catalyst. The effects of methanol to oil molar ratio, dosage of enzyme and reaction time were investigated. The optimum reaction conditions for fresh enzyme were methanol to oil molar ratio of 25:1, 10 % of Novozym 435 based on oil weight and reaction period of 4 h at 50 °C obtaining a biodiesel yield of 89.1 %. Moreover, the reusability of the lipase over repeated cycles was also investigated under standard conditions.

Tamalampudi et al. (2008) employed immobilized whole cell and commercial lipase as biocatalyst for biodiesel production from Jatropha oil. The lipase producing whole cells of Rhizopus oryzae (ROL) immobilized onto biomass support particles (BSPs) were used for the production of biodiesel from relatively low-cost nonedible oil from the seeds of Jatropha curcas. The activity of ROL was compared with that of the commercially available, most effective lipase (Novozym 435).

The various alcohols were tested as a possible hydroxyl donor, and methanolysis of jatropha oil progresses faster than other alcoholysis regardless of the lipases used. The maximum methyl esters content in the reaction mixture reaches 80 wt% after 60 h using ROL, whereas it is 76 % after 90 h using Novozym 435. Zheng, Wua, Christopher, Jing, and Zhu (2009) reported the lipase catalyzed transesterification process in a solvent-free system. Feruloylated diacylglycerol (FDAG) was synthesized using a selective lipase-catalyzed transesterification between ethyl ferulate and triolein. The highest reaction conversion and selectivity toward FDAG were 73.9 and 92.3 %, respectively, at 338 K, reaction time of 5.3 days, with enzyme loading of 30.4 mg/ mL; water activity is 0.08, and the substrate molar ratio is 3.7. The disadvantage of the enzyme-catalyzed process is that it is time consuming compared to acid or base catalyzed transesterification. Enzymes have several advantages over chemical catalysts such as mild reaction conditions; specificity, reuse; and enzymes or whole cells can be immobilized, and are considered natural, and the reactions they catalyze are considered green reactions (Akoh et al., 2007). However, the drawbacks of enzymatic catalysts are significantly higher production cost (Meher et al., 2006) as well as difficulty during manufacturing due to the need for a careful control of reaction parameters (Cervero, Coca, & Luque, 2008). Moreover, the reaction yields as well as the reaction times are still unfavorable compared to the alkaline catalyzed reaction systems.

2.2.5 Catalyst Support/Carrier for Heterogeneous Catalysis

One of the ways to minimize the mass transfer limitation for heterogeneous catalysts in liquid phase reactions is using catalyst supports. The choice of a particular
support depends on the nature of application and reaction conditions. In general, the
support materials should have a high surface area and suitable mechanical strength
to permit dispersion of the metal, also to increase its thermal stability and hence the
catalyst life. The common support materials have been used in biodiesel production
are presented in Table 2.3.

The selection of a support is based on it having certain properties, some of which
are listed below (Chuah, Jaenicke, & Xu, 2000; Perego & Villa, 1997):

- Inertness. Ideally, support materials should have no catalytic activity leading to
  undesirable side reactions.
- Desirable mechanical properties, including attrition resistance, hardness, and
  compressive strength.
- Stability under reaction and regeneration conditions.
- Surface area. High specific surface area is usually, but not always, desirable.
- Porosity, including average pore size and pore size distribution. High area implies
  fine pores, but relatively small pores could become plugged during impregna-
  tion, especially if high loading is sought.
- Stability at high temperature reactions
- Low cost

Materials with low surface area are generally useful in supporting very active
catalytic components in reactions where further side reactions may affect the activity
and selectivity, while materials with high surface areas are widely used in precious
metal catalysts preparation (Perego & Villa, 1997). In addition, the catalytic behavior
of the supported particles depends strongly on their interactions with the support.
The active species in catalysts can be either in the metallic (e.g., iron, copper, and

Table 2.3 Common carriers
employed for
transesterification reaction

Support/carrier	Chemical formula
Alumina	γ-Al_2O_3
Silica	SiO_2
Zirconia	ZrO_2
Titania	TiO_2
Ceria	CeO_2
Magnesium oxide	MgO
Activated carbon	C
Tungsten oxide	WO_3
Silica carbide	SiC
Zeolites	NaX
Stannous oxide	SnO
Zinc oxide	ZnO

platinum) or metal oxide form or other forms (e.g., sulfides and carbides). Among the support materials, alumina (γ-Al$_2$O$_3$) has received considerable attention for catalytic reactions as a support owing to its high surface area and stability at high temperature reactions (>800 °C) (Chuah et al., 2000).

2.2.6 Production of Alumina (γ-Al$_2$O$_3$) Support Particle

An approach for the preparation of a macroscopic support has been developed by Prouzet, Tokumoto, and Krivaya (2004). According to the terminology proposed by Prouzet et al. (2004), this method corresponds to a "integrated gelling" procedure. The two basic components involved in the present gelation process are sodium alginate and boehmite, where the gelation of boehmite under sol–gel process and biopolymer (sodium alginate) under ion exchange process. This process has been applied to form ceramic supports, where sodium alginate is dispersed drop wise from a needle into a magnetically stirred solution of aluminum chloride hexahydrate (AlCl$_3$.6H$_2$O) to form rubbery, substantially hard supports. One of the most important contributions in the field were made by Backov (2006), who synthesized millimetric range of beads from combined gelation of sol–gel and ion exchange process. In a similar approach, macroscopic support (2 mm) was synthesized from boehmite sol doped with sodium alginate, followed by calcination to give an oxide phase (Prouzet ct al., 2006).

Another promising synthesis route suitable to produce particles is by combining sol-gel and oil drop dripping technique. Sol-gel, which involves the formation of a colloidal solution followed by formation of a gel, typically uses colloidal dispersions as the starting material. This process has been applied to form milli-sized supports, in which a aqueous gel is pressed through a syringe with a constant flow rate and droplets are formed at the tip of the needle. Subsequently, the droplets are dripped into a column of paraffin oil and ammonia solution, followed by drying and calcining. Applying this technique, the preparation of copper oxide coated gamma alumina granules with a diameter of about 2 mm has been described by Wang and Lin (1998). A similar method has been used, sometime after, by Buelna and Lin (1999) and has emphasized the influence of drying technique (microwave technique) on the properties of copper oxide coated alumina granules. A method for continuous preparation of granular particles has been developed by the same research group (Buelna & Lin, 2004) and attempted to address the characterization and desulfurization-regeneration properties of the alumina supported copper oxide granule.

In comparison, the integrative method is more simple and easy to scale up than the oil-drop granulation method. This is because the second method requires the use of oil that adds to complication in the design of bead collection system and requires vigorous rinsing of beads to get rid of oil at the bead surface. Therefore, comparison of the particle properties produced from two methods was made to select the suitable method in order for it to be used subsequent works.

2.2.7 *Production of Supported Catalyst*

The aim of the preparation of catalytic materials is to prepare a product with high activity, selectivity, and stability (Pinna, 1998). Although some catalytic materials are composed of single substances, most catalysts have different types of easily distinguishable components, active components, a support (Richardson, 1989). Figure 2.4 illustrates the various steps carried out for the preparation of supported catalysts using preshaped catalytic supports.

The usual pathway (1, 3, 4, and presumably 5) involves impregnation of the support grains or pellets with an aqueous solution containing the species of the active element to be deposited followed by drying at low temperatures (20–100 °C) and then calcination at high temperatures (usually 300–1000 °C).

Supports can provide higher surface area through the existence of pores where metal particles can be anchored (Chorkendorff & Niemantsverdriet, 2003). The active components are responsible for the principal chemical reaction. The active metal component is usually deposited on the surface of a porous or non support as shown in Fig. 2.5. The different methods which are usually used to prepare supported catalysts will be discussed in subsequent sections.

2.2.7.1 Impregnation

Impregnation is a procedure whereby a certain volume of solution is contacted with the solid support, which, in a subsequent step, is dried to remove the imbibed solvent. Two methods of impregnation may be distinguished, depending on the volume of

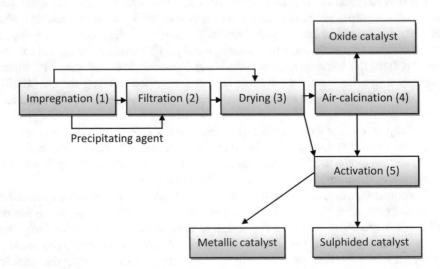

Fig. 2.4 General preparation scheme of supported catalyst. Source: Bourikas, Kordulis, and Lycourghiotis (2006)

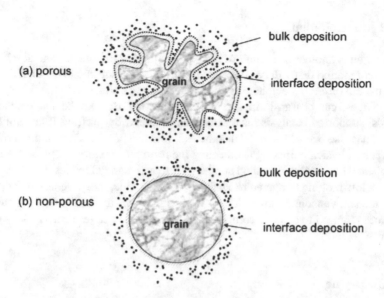

Fig. 2.5 Schematic representation of the supported catalytic particles; (**a**) porous support, (**b**) non-porous support. Source: Bourikas et al. (2006)

solution used: incipient wetness or dry impregnation and wet or soaking impregnation (Mul & Moulijn, 2005, Richardson, 1989). In incipient wetness impregnation method, the volume of the solution is equal or slightly less than the pore volume of the support. The volume should be just sufficient to fill the pores and wet the outside of the particles. Although this volume may be determined from measured pore volumes, this is sometimes more reliably determined with preliminary test on aliquot samples (Richardson, 1989).

The maximum loading is limited by the solubility of the precursor in the solution (Pinna, 1998). When higher loadings are required, this limitation is overcome by carrying out consecutive impregnation steps. In wet impregnation an excess of solution with respect to the pore volume of the support is used (Campanati, Fornasari, & Vaccari, 2003, Schwarz, Contescu, & Contescu, 1995). The system is left to age for a certain period under stirring it is then filtered and dried. This procedure is applied when a precursor support interaction can be envisaged. Therefore, the concentration of the support will not only depend on the concentration of the solution and on the pore volume of the solution, but also on the type and/or concentration of the adsorbing sites existing at the surface. In general, wet impregnation is used for the preparation of low-loaded catalysts and in particular expensive precious metal catalysts, where the active metal phase should be highly dispersed in order to obtain high activity. The distribution of the metal precursor will be based on the density of the exchanging sites in the support. With low metal loading and high density of adsorbing sites on supports in granules, pellets, extrudates (where diffusion effects are encountered), the distribution of the precursor will be inhomogeneous (Pinna, 1998). Deposition will mainly take place at the external layers of the particles.

2.2.7.2   Precipitation

Precipitation is another procedure where the solutions containing the metal salt and a salt of a compound that will be converted into the support are contacted under stirring with a base in order to precipitate as hydroxide and or/carbonates (Pinna, 1998). After washing, these can be transformed to oxides by heating. Besides the abovementioned methods, supported catalysts have also been prepared by grafting (Schwarz et al., 1995). It has also been discovered that metal oxides can be deposited on the surface of supports by physically mixing and heating the resulting mixture to spread the active component. However, this method applies only to active metal oxides that are volatile or have a low melting temperature, such as rhenium oxide, molybdenum oxide, tungsten oxide and vanadium oxide. The disadvantage of this method is the long calcination times required to achieve complete spreading of the active metal oxide over the support surface.

2.2.7.3   Drying

Drying is an important step in catalyst preparation since it can affect the distribution of the active species. During drying, the solution in the pores will become oversaturated and precipitation takes place (Mul & Moulijn, 2005). If not done properly, this step can result in irregular and uneven concentration distributions (Richardson, 1989). Different variables such as the heating rate, final temperature and time of treatment and type of atmosphere can influence the drying process and have to be selected according to different systems (Pinna, 1998). In principle, rapid evaporation of the solvent is favorable because it causes rapid supersaturation of the solution in the pores and that is associated with a high dispersion of the active species (Mul & Moulijn, 2005). However, if the drying rate is too slow evaporation occurs at the meniscus, which retreats down the pore, some salt deposition occurs but most of the solute merely concentrates deeper in the pore (Pinna, 1998, Richardson, 1989). When finally crystallized, the salt is located at the bottom of a pore or at the particle center (Pinna, 1998).

2.2.7.4   Calcination and Activation

Calcination is a thermal treatment process applied solid materials to bring about a thermal decomposition or removal of a volatile fraction. The calcination process normally takes place at temperatures below the melting point of the product materials (Pinna, 1998). The purpose of calcination is to decompose the metal precursor with the formation of an oxide and removal of the cations or the anion that have been previously introduced as gaseous products. Besides decomposition, during the calcination sintering of the precursor or of the formed oxide or a reaction of the formed oxide with the support can occur (Pinna, 1998, Richardson, 1989). When dealing with bimetallic catalysts, control of calcination temperature is required in order to avoid the formation of two separate oxides or segregation of one of the components.

Other thermal treatments, such as reduction or sulfidation, which are performed in a special atmosphere, are called activation operations (Perego & Villa, 1997). Variables such as the rate of heating, the time of calcination have to be carefully chosen depending on the type of metal, catalytic system, and reaction type (Pinna, 1998).

2.2.8 Supported-Catalysts Employed for Biodiesel Production

Heterogeneously catalyzed transesterification reaction is complex because it occurs in a three-phase system consisting of a solid (heterogeneous catalyst) and two immiscible liquid phases (oil and methanol). The need for development of heterogeneous catalysts has arisen from the fact that homogeneous catalysts used for biodiesel development pose a few drawbacks discussed in previous sections. Easy separation, easy recovery, no problems in solubility and miscibility are the strengths of a heterogeneous system in order to reduce the cost of production. Heterogeneous catalysis is thus considered to be a green process. Needless to say, because of these advantages, research on the transesterification reaction using heterogeneous catalysts for biodiesel production has increased over the past decade.

A great variety of materials have been tested as heterogeneous catalysts for the transesterification of vegetable oils, as shown in Table 2.4. There are number of process variables which could affect the transesterification process. In addition, the catalyst efficiency depends on several factors such as specific surface area, pore size, pore volume, acidity or basicity, and active site concentration of catalyst (Smith & Notheisz, 2006) which are discussed in subsequent sections.

Table 2.4 Heterogeneous catalysts employed for transesterification reactions

Catalyst type		Example
Solid acid catalyst	Zeolite type solid acid catalyst	Zeolite Socony Mobil-5 (HZSM-5); zeolite-β; zeolites-Y
	Heteropoly Acid Loaded MCM-41 Catalyst	Mg-mobile crystalline material-41(MCM-41); Al-MCM-41
	Sulfated zirconia and tin oxide type solid acid catalyst	SO_4^{2-}/ZrO_2; SO_4^{2-}/SnO_2
	Tungsten trioxide loaded zirconia type solid acid catalyst	WO_3/ZrO_2
Solid basic catalyst	Alkali metal salts loaded on alumina	KI/Al_2O_3; $Mg(NO_3)_2/Al_2O_3$ $Na/\gamma-Al_2O_3$; $Na/$ $NaOH/\gamma-Al2O3$; $NaOH/\gamma-Al_2O_3$; KF/Al_2O_3; $KCO_3/$ Al_2O_3; KNO_3/Al_2O_3; $LiNO_3/Al_2O_3$; $Ca(NO_3)_2/Al_2O_3$; $NaNO_3/Al_2O_3$; KOH/Al_2O_3
	Alkaline earth metal oxide	MgO; CaO; $CaCO_3$; $Ca(OH)_2$ SrO; $CaO–La_2O_3$,
	Hydrotalcites	MgO/Al_2O_3; Mgo/CaO
	Zeolites	ETS-10; KOH/NaX; NaO/NaX; CsX; KI/NaX

2.2.9   Preparation of Supported Materials

Boehmite suspensions find wide applications as precursor materials for the preparation of alumina and alumina-based powders, catalysts support material, adsorbents (Cheng, Lan, Feng, Zhan-Wen, & Fei, 2006; Kirkland, 1963; Liu et al., 2008). The understanding on the rheological properties of the suspension is important for process control and optimization. Cristiani, Valentini, Merazzi, Neglia, and Forzatti (2005) and Tsai and Yung (2007) studied the effect of aging time on rheology of alumina slurries and Fauchadour, Kolenda, Rouleau, Barre, and Normand (2000) studied the peptization mechanism of boehmite suspension. Song and Chung (1989) studied the rheological properties of aluminum alkoxide in acid and basic solutions. In a similar approach, Moreno, Salomoni, and Stamenkovic (1997) and Schilling, Sikora, Tomasik, Li, and Garcia (2002) studied the effect of additives on the rheology of alumina slurry; until now, there is little information on the rheological properties of boehmite suspensions as a function of process variables such as pH, temperature, and concentration.

2.2.10   Effect of Process Variables

Optimization of these process variables is usually carried out by changing the certain variables while maintaining the other variables at a fixed values and then subsequently comparing the yield or conversion of fatty acid alkyl esters (FAME). These factors or variables usually have different effect on the transesterification process depending on the catalyst used for the transesterification process.

2.2.10.1   Effect of Molar Ratio of Alcohol to Oil

One of the most important factors that affect the yield of ester (biodiesel) is the molar ratio of alcohol to triglyceride (oil). Based on the stoichiometric of transesterification reaction, every mol of triglyceride requires three moles of alcohol to produce three moles of fatty acid alkyl esters and one mole of glycerol. Transesterification is an equilibrium reaction in which an excess of alcohol is required to drive the reaction to the right (Ma & Hanna, 1999). However, an excessive amount of alcohol makes the recovery of the glycerol difficult, so that the ideal alcohol–oil ratio has to be established empirically (Schuchardt, Serchelia, & Vargas, 1998).

Transesterification of rapeseed oil carried out by Kawashima et al., (2008) showed a maximum conversion at 6:1 of methanol to oil ratio, whereas an earlier study by Xie and Huang (2006) found a maximum conversion at a ratio of 15:1. With further increase in molar ratio the conversion efficiency more or less remains the same. Most of the studies on the solid base-catalyzed transesterification of vegetable reported that maximum conversion to the ester occurred with a molar ratio of 6:1 (Kim et al., 2004; Kawashima et al., 2008; Portnoff, Purta, Nasta,

Zhang, & Pourarian, 2006). However, the other results showed that the optimum molar ratio of oil to alcohol was 9:1 (Modica et al., 2004), 40:1 (Tateno & Sasaki, 2004), 12:1 (Albuquerque et al. (2008), 15:1 (Xie, Peng, & Chen, 2006a) and 30:1 (Ngamcharussrivichai, Totarat, & Bunyakiat, 2008) to get the maximum biodiesel yield. Solid acid-catalyzed reactions require the use of high alcohol-to-oil molar ratios in order to obtain good product yields in practical reaction times. Higher molar ratios showed only moderate improvement until reaching a maximum value at a 55:1 ratio (82 %) (Antunes, Veloso, Assumpc, & Henriques, 2008).

Therefore, higher molar ratio of oil to alcohol (>6:1) could also be used as the optimum ratio for oil to methanol, depending on the quality of feedstock and catalyst type of the transesterification process.

2.2.10.2   Effect of Catalyst Amount

The type and amount of catalyst required in the transesterification process usually depend on the quality of the feedstock applied for the transesterification process. For a purified feedstock, any type of catalyst could be used for the transesterification process. However, for feedstock with high moisture and free fatty acids contents, homogenous transesterification process is unsuitable due to high possibility of saponification process instead of transesterification process (Gerpen, 2005). The yield of fatty acid alkyl esters generally increases with increased amount of catalyst (Demirbas, 2007; Ma & Hanna, 1999). This is due to availability of more active sites by additions of larger amount of catalyst in the transesterification process. However, the addition of an excessive amount of catalyst, however, gives rise to the formation of an emulsion, which increases the viscosity and leads to the formation of gels (Encinar, Gonzalez, & Rodriguez-Reinares, 2005). These hinder the glycerol separation and, hence, reduce the apparent ester yield. Therefore, similar to the ratio of oil to alcohol, optimization process is necessary to determine the optimum amount of catalyst required in the transesterification process.

In the case of the heterogeneous catalysis, the literature presents many works relating to this issue. In most of the literature reviewed the results showed that the best suited catalyst concentrations giving the best yields of the esters are between 2.5 and 10 wt% (Benjapornkulaphong, Ngamcharussrivichai, & Bunyakiat, 2009; Boz, Degirmenbasi, & Kalyon, 2009; Noiroj, Intarapong, Luengnaruemitchai, & Jai-In, 2009; Samart, Sreetongkittikul, & Sookman, 2009). Xu, Li, Hu, Yang, and Guo (2009) used Ta_2O_5/SiO_2-$[H_3PW_{12}O_{40}/R]$ (R = Me or Ph) as the catalyst in the transesterification of soybean oil and reported that 2 wt% (in terms of oil) catalyst is the optimum catalyst concentration. Similarly, Xie, Yang, and Chun (2007) carried out transesterification of waste rapeseed oil and obtained maximum conversion at 3 wt% NaX/KOH catalyst concentration and further increase in the catalyst concentration had no effect on conversion.

These results were confirmed by Xie et al. (2006a) who carried out transesterification of soybean oil with KI/Al_2O_3 concentrations at 0.05 wt% increments starting from 1 to 3.5 wt% and observed that the highest conversion was achieved at 2.5 wt% concentration.

2.2.10.3   Effect of Reaction Time

For a heterogeneous transesterification process, the reaction period varies depending on the reactivity and type of the solid catalyst used. For a practical and economic feasible transesterification process, it is necessary to limit the reaction time at a certain period. Longer reaction time could also permit reversible transesterification reaction to occur, which eventually could reduce the yield of fatty acid alkyl esters (Demirbas, 2009). Thus, optimization of reaction time is also necessary.

Most investigators have observed an optimum reaction time for basic-catalyzed transesterification process around 3–12 h (Aderemi & Hameed, 2009; Boz et al., 2009; Benjapornkulaphong et al., 2009; Kolaczkowski, Asli, & Davidson, 2009; Lukic, Krstic, Jovanovi, & Skala, 2009; Noiroj et al., 2009). However, a direct comparison of the biodiesel yield with the reaction time is difficult because of the variation in other reaction conditions, notably, the oil to methanol molar ratio and amount of catalyst used. Current researches have shown that the reaction time for a non-catalytic transesterification process using supercritical alcohol is shorter compared to conventional catalytic transesterification process (Demirbas, 2003, 2007). However, non-catalytic transesterification process using supercritical alcohol is much more energy intensive than the solid base-catalyzed process. Because it operates at very high pressures (200–450 bar) and the high temperatures (350–400 °C) bring along proportionally high heating and cooling costs. It was also reported that excess reaction time does not increase the conversion but favors the backward reaction (hydrolysis of esters) which results in a reduction of product yield (Leung & Guo, 2006).

2.2.10.4   Effect of Reaction Temperature

Temperature is an important parameter as it allows the faster reaction kinetics and mass transfer rates in the transesterification reaction (Corma, 1997; Freedman & Pryde, 1984; Liu, 1994). Normally, a relatively high reaction temperature is required for heterogeneous system in order to increase the mass transfer rate between reactant molecules and catalyst. This is due to the existence of initial three-phase mixture; oil–methanol–solid catalyst. Higher temperatures decrease the time required to reach maximum conversion (Pinto et al., 2005).

Many studies have shown that reaction temperature significantly influences FAME yield for transesterification reaction catalyzed by heterogeneous process. Boz et al. (2009) found that the yield of biodiesel was tripled (30–99 %) using solid base catalyst (KF/γ-Al$_2$O$_3$) when the transesterification temperature increased from 25 to 65 °C. Similar results were reported by Samart et al. (2009). The conversion was increased from 68 % at 50 °C to 90 % at 70 °C with a 16:1 M ratio of methanol to oil using KI/mesoporous silica catalyst. The influence of FAME yield on TiO$_2$/SiO$_2$ as a solid acid-catalyzed transesterification has been studied by Serio et al. (2007). Their study shows that the yield of FAME was increased from 5 to 62 % when the temperature was changed from 120 °C to 180 °C for transesterification of soybean oil with molar ratio of methanol to oil of 1/1 (w/w). Similarly, Lam and Lee (2010) showed that transesterification yield of waste cooking oil increased from 80.3 to 88.2 % as the temperature increased from 60 to 100 °C. The methanol to oil molar ratio was 15, the catalyst (SO^{-4}/SnO$_2$–SiO$_2$) (referring to weight of oil) was 6 wt% and 1 h reaction time was used.

Researchers have found that that the reactions are accelerated at critical point conditions. The critical temperatures and critical pressures of the various alcohols are shown in Table 2.1. Madras et al. (2004) showed that transesterification conversion of sunflower oil increased from 78 to 96 % as the temperature increased from 200 to 400 °C. The methanol to oil molar ratio was 40, the pressure was 200 bar, and a 40 min reaction time was used. Similar results were reported by Demirbas (2002).

The conversion at 5 min can be nearly doubled from 50 % at 177 °C to over 95 % at 250 °C with a 41:1 molar ratio of methanol to hazelnut kernel oil. Other results consistent with this finding are from Demirbas (2003). Their study shows that that the yield of FAME increased (from 5 to 99 %) when the transesterification temperature increased from 127 to 337 °C as shown in Fig. 2.6. Thus, the temperature had a favorable effect on fatty acid methyl ester (FAME) yield.

Transesterification can be conducted using solid catalyst at various temperatures ranging from 60 to 450 °C. However, the operating temperature for transesterification process depends on the method used. Certain processes, heterogeneous acid catalyzed reaction transesterification process, generally require moderate temperature ranging from 120 to 250 °C (Almeida, Noda, Goncalves, Meneghetti, & Meneghetti, 2008; Furuta, Matsuhashi, & Arata, 2004; Garcia, Teixeira, Marciniuk, & Schuchardt, 2008).

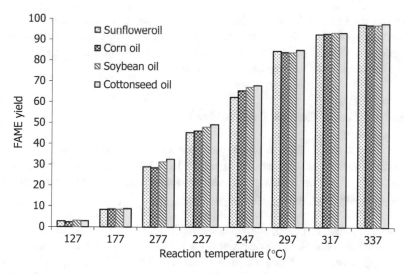

Fig. 2.6 Changes in yield percentage of methyl esters as treated with supercritical methanol at different temperatures as a function of reaction temperature. Source: Demirbas (2003)

However, non-catalytic transesterification process requires high temperature ranging from 230 to 450 °C to yield the desired product (fatty acid alkyl esters) (Demirbas, 2008; He, Sun, et al., 2007a, 2007b; Saka & Dadan, 2001). A great variety of solid basic catalysts such as alkaline-earth metals oxides and hydroxides, alkali metals hydroxides or salts supported on alumina, zeolites, hydrotalcites have been evaluated to date which have shown to be good candidates for transesterification reaction at relatively low temperatures ranges (60–70 °C) (Bo, Guomin, Lingfeng, Ruiping, & Lijing, 2007; Liu, He, Wang, & Zhu, 2007); MacLeod, Harvey, Lee, & Wilson, 2008.

2.2.10.5   Effect of Mixing Intensity

Mixing is very important in the transesterification process, as oils or fats are immiscible with alcohol. As a result, vigorous mixing is required to increase the area of contact between the two immiscible phases (Meher, et al., 2006; Singh & Fernando, 2006a; Singh et al., 2006). Mechanical mixing is commonly used in the transesterification process. The intensity of the mixing could be varied depending on its necessity in the transesterification process. In general, the mixing intensity must be increased to ensure good and uniform mixing of the feedstock. When vegetable oils with high kinematic viscosity are used as the feedstock, intensive mechanical mixing is required to overcome the negative effect of viscosity to the mass transfer between oil, alcohol and catalyst.

Poor mass transfer between two phases in the initial phase of the reaction results in a slow reaction rate, the reaction being mass transfer controlled (Noureddini & Zhu, 1997). Stamenkovic, Lazic, Todorovic, Veljkovic, and Skala (2007) studied the effect of agitation intensity on alkali-catalyzed methanolysis of sunflower oil and reported that the drop size distributions of emulsion were found to become narrower and shift to smaller sizes with increasing agitation speed. It is evident, from the literature presented above, that the agitation had a favorable effect on fatty acid methyl ester (FAME) yield. Therefore, variations in mixing intensity are expected to alter the kinetics of the transesterification reaction.

2.2.10.6   Effect of Catalyst Particle Size

The catalytic activity of the materials has been reported to be dependent on the particle size. Smaller particles can be expected to exhibit higher rate of reaction, or consequently conversions for a given volume of reaction mass due to increased external surface available (McCarty & Weiss, 1999).

Gutierrez-Ortiz, Lopez-Fonseca, Gonzalez Ortiz de Elguea, Gonzalez- Marcos, and Gonzalez-Velasco (2000) studied in details the catalytic hydrogenation of methyl oleate, which is the main component of olive oil, using a Ni/SiO$_2$ catalyst in a slurry reactor. The authors concluded that, at 6 bars and 180 °C, both activity and selectivity significantly decreased when the size of the catalyst particles were larger than 50 µm and the stirring rate was below 2000 rpm. Ensoz, Angın, and Yorgun (2000) investigated the influence of particle size on the pyrolysis of rapeseed by varying the particle size of rapeseed in the range of 0.224–1.8 mm and found that the yields of products are largely independent of particle size. Other results consistent with these findings are from Ferretti, Olcese, Apesteguia, and Di Cosimo (2009). In order to investigate the effect of the MgO particle size on FAME conversion, they carried out several 3-h catalytic tests using three different particle size ranges (100, 100–177, and 177–250 μm), without changing any other reaction parameter. Only small differences were observed during the 3-h tests. The two catalytic tests with the smallest particles show slightly lower FAME conversions than the experiment with the largest size. This result is the opposite of that expected in the presence of diffusional limitations. The authors concluded that this effect could be attributed to a "flotation effect" of the smallest particles in the presence of the foam caused by the surfactant monoglyceride that probably places the catalyst surface far from the glycerol phase, thereby decreasing the FAME conversion. Therefore, for practical reasons and to avoid flotation of small particles, the largest particle size range has been adopted for the glycerolysis of fatty acid ethyl esters using MgO catalyst.

Recent advances in nanoscience and nanotechnology have led to a new research interest in employing nanometer-sized particles as an alternative matrix for supporting catalytic reactions. Compared with conventional supports like solid-phase, nanoparticular matrices could have a higher catalyst loading capacity due to their very large surface areas (McCarty & Weiss, 1999). Freese, Heinrich, and Roessner (1999) have reported the catalytic activity of micrometer zeolite sieve of molecular porosity (ZSM-5) catalysts and found that the catalyst exhibited a conversion of 14.2 %, which is much lower than the nanocrystalline ZSM-5 catalysts used in their investigation. The higher activity of nanocrystalline ZSM-5 is probably due to the increased external surface of smaller crystal.

Mabaso, Van Steen, and Claeys (2006) reported the effect of crystal size on carbon supported iron catalysts prepared via precipitation where catalysts with smaller metal than 7–9 nm have showed higher selectivity of methane conversion compared to the bigger-sized catalysts. However, the effect of particle size on fatty acid methyl ester has been studied to a lesser extent. Recently the utilization of Al_2O_3 supported KF catalyst having a particle size in the order of nanometer order for biodiesel production has been demonstrated by Boz et al. (2009) and reported that the catalyst can be used in the production of biodiesel from vegetable oil. The high yield of fatty acid methyl ester (FAME) (>90 %) has been achieved using the catalyst ranging from nanometer to micrometer in diameter. However, from the practical point of view, handling of such small particles in large quantities could be difficult due to the formation of dust.

The smaller the emitted particle, the more harmful it is to the human body because particles under 100 nm (ultrafine particles) in diameter have a higher surface area per unit mass of particles; therefore, the smaller particles can more easily infiltrate into the respiratory organs (Donaldson, Li, & MacNee, 1998). Utilization of powders in conventional catalytic reactions is problematic because powder form catalysts are at a disadvantage in pressure drop, mass/heat transfer, contacting efficiency, and separation processes (Centi & Perathoner, 2003a, 2003b; Matatov & Sheintuch, 2002). Therefore, the design of a catalyst's form at a macroscale is indispensable to avoid these problems. From these viewpoints, macrostructured materials have drawn attention as catalytic supporting materials (Centi & Perathoner, 2003a; Matatov & Sheintuch, 2002). It was reported by Nabeel, Jarrah, Ommen, and Lefferts (2004) that the macroscopic particle will open-up a real opportunity for their use as a catalyst support in relation to the traditional catalysts carriers. Among the different potential applications of these materials, catalysis either within the gas or the liquid phase seems to be the most promising according to the results recently reported in literature (Nabeel et al., 2004).

2.2.10.7   Shape of Particles

The shape of the particles was measured quantitatively by means of sphericity factor (SF) as described by Chan, Lee, Ravindra, and Poncelet (2009). The sphericity factor provides brief classification about the degree of deviation of the irregular particle from the true sphere shape with 0 as perfect sphere and increasing value indicates higher degree of deformation. It has been shown by Chan et al. (2009) that a particle with SF less than 0.05 can be considered as spherical. In addition, supported catalyst in spherical form can offer shape-dependent advantages such as minimizing the abrasion of catalyst in the reaction environment as reported by Campanati et al. (2003). In fact, the stability of the catalyst will be increased.

2.2.11   Effect of Catalyst Properties

Synthesizing novel heterogeneous catalysts that have desirable physical and chemical properties for biodiesel production is one of the focuses of the latest research. The catalyst efficiency depends on several factors such as specific surface area, pore size, pore volume, acidity or basicity, and active site concentration of catalyst (Smith & Notheisz, 2006) which will be discussed in subsequent sections.

2.2.11.1   Effect of a Base/Acid Catalyzed Reaction

The density of acidic or basic sites is important in determining a solid catalyst's activity and selectivity; however, it is not easy to evaluate the relative importance of the two types of sites in the reaction (Serio, Tesser, Pengmei, & Santacesaria, 2008; Xie et al., 2007). Solid, base catalysts have a higher activity and faster reaction rate compared to solid, acid catalysts (Lotero et al., 2006). Solid, base catalysts are very sensitive to the presence of water and FFA. In other words, they need feedstock with a low FFA to avoid deactivation of the catalyst. In contrast to solid bases, solid, acid catalysts can be applied to feedstock with a high FFA and water content. Unfortunately, solid, acid catalysts need higher temperatures and catalyst loadings to obtain reasonable FAME yield (Schuchardt et al., 1998).

2.2.11.2 Solid Basic Catalysts

A wide variety of solid, basic catalysts, such as alkaline-earth metals oxides and hydroxides and alkali metals, hydroxides and salts supported on alumina, zeolites and hydrotalcites, have been studied with different reaction conditions; they are summarized in Table 2.5. KI, KF, and KNO_3 catalysts supported on alumina showed good activities at low temperatures due to their basic sites forming either K_2O species produced by thermal decomposition or Al–O–K groups formed by salt–support interactions (Xie et al., 2006a; Xie, Peng, & Chen, 2006b; Vyas, Subrahmanyam, & Patal, 2009). In a similar approach, the catalytic activities of $Na/NaOH/\gamma-Al_2O_3$ and $K/KOH/\gamma-Al_2O_3$ towards transesterification were promoted by the presence of strong, basic sites in the catalysts that originated by the ionization of sodium or potassium (Kim et al., 2004; Ma, Li, Wang, Wang, & Tian, 2008). Unfortunately, leaching of the sodium or potassium during the transesterification reactions indicated a lack of chemical stability for the catalysts under the reaction conditions.

More than 90 % of biodiesel yield was obtained using $Ca(NO_3)_2/Al_2O_3$ and $LiNO_3/Al_2O_3$, while $Mg(NO_3)_2/Al_2O_3$ and $NaNO_3/Al_2O_3$ catalysts possessed an inactive magnesium and sodium-aluminate phase, resulting in low biodiesel yield (Benjapornkulaphong et al., 2009). Additionally, a small amount of Li ion was found to leach out from the $LiNO_3/Al_2O_3$ catalysts during the reaction. Therefore, $Ca(NO_3)_2/Al_2O_3$ seemed to be a promising catalyst due to its high conversion (90 %) in the transesterification reaction and its stability under the reaction conditions (Benjapornkulaphong et al., 2009).

Many research studies on transesterification reactions catalyzed by alkaline earth oxides have also been reported in literature. Calcium, Mg, Sr, and Ba oxides have been all treated at high calcination temperatures (500–1050 °C). In particular, CaO was found to be active for the transesterification of vegetable oil in refluxing methanol (Demirbas, 2007; Liu, He, Wang, Zhu, & Piao, 2008b; Kouzu et al., 2008). It was proposed that CaO catalyzed the transesterification reaction through a nucleophilic reaction and accelerated by the enhancement of the basic properties (Kouzu et al., 2008), as shown in Fig. 2.7.

The sequence of steps can be summarized as follows. In the first step, the CaO reacted with the alcohol, producing an alkoxide (RO⁻) and the protonated (H⁺) catalyst. Nucleophilic attack of the alkoxide at the carbonyl group of the triglyceride then generated a tetrahedral intermediate (Lotero et al., 2006) from which the alkyl ester and the corresponding anion of the diglyceride were formed. The latter deprotonated the catalyst to regenerate the active species that was able to react with a second molecule of alcohol to start another catalytic cycle. Lastly, the monoglycerides were converted using the same mechanism to a mixture of alkyl esters and glycerol. In such a catalyst–substrate interaction, the electrophilicity of the adjacent carbonyl carbon atom increases, making it more susceptible to nucleophilic attack. Similarly, this dependency has been reported for the $Ca(OH)_2$, $CaCO_3$, MgO, and $Ba(OH)_2$ catalysts for their transesterification reactions (Antunes et al., 2008; Leclercq, Finiels, & Moreau, 2001; Tateno & Sasaki, 2004). However, the activation of the base sites in MgO was found to be greater at higher calcination temperatures (Leclercq et al., 2001).

Table 2.5 FAME yield using basic catalysts in the transesterification of vegetables oil

Catalysts	M/O	oil	RT (°C)	MFY	MFP	FY (%)	Catalyst type	References
KI/Al$_2$O$_3$	30:1	Palm	MRT	0.05	0.14	87.4	Alkali metal salts loaded on alumina	Xie et al., 2006a
Mg(NO$_3$)$_2$/Al$_2$O$_3$	65:1	Palm	60	0.01	0.03	10.4		Benjapornkulaphong et al., 2009
Na/γ-Al$_2$O$_3$[a]	6:1	Soybean	60	0.04	0.35	70.0		Kim et al., 2004
Na/NaOH/γ-Al$_2$O$_3$[a]	6:1	Soybean	60	0.04	0.39	78.0		Kim et al., 2004
NaOH/γ-Al$_2$O$_3$[a]	6:1	Soybean	60	0.03	0.33	65.0		Kim et al., 2004
KF/Al$_2$O$_3$	15:1	Soybean	MR	0.04	0.14	85.8		Xie et al., 2006a
KCO$_3$/Al$_2$O$_3$	15:1	Soybean	MR	0.02	0.08	48.0		Xie et al., 2006a
KNO$_3$/Al$_2$O$_3$	12:1	Jatropha	70	0.02	0.10	84		Vyas et al., 2009
NaOH/Al$_2$O$_3$	12:1	Sunflower	50	0.50	0.03	88		Arzamendi et al., 2007
KNO$_3$/Al$_2$O$_3$	15:1	Soybean	MR	0.02	0.16	87		Xie et al., 2006b
MgO	55:1	Soybean	130	0.02	0.12	82	Alkali earth oxide based catalyst	Antunes et al., 2008
CaO[b]	6:1	Sunflower	252	0.02	2.60	65		Demirbas, 2007
CaCO$_3$[b]	39.3:1	Soybean	250	0.08	5.43	87		Tateno & Sasaki, 2004
Ca(OH)$_2$[b]	39.4:1	Soybean	300	0.16	6.13	98		Tateno & Sasaki, 2004
MgO[b]	39.6:1	Soybean	300	0.08	5.69	91		Tateno & Sasaki, 2004
Sodium silicate[c]	6:1	Castor oil	120	0.05	4.37	70		Portnoff et al., 2006
SrO	12:1	Soybean	70	0.04	1.90	95		Liu et al., 2007
Ca(OCH$_3$)$_2$	1:1	Soybean	65	0.05	0.32	98		Liu et al., 2008d
Ba(OH)$_2$[c]	9:1	Rapeseed	MR	0.20	3.91	97		Mazzocchia, Modica, Kaddouri, & Nannicini, 2004

Catalyst	M/O	Oil	MR			FAME %	Category	Reference
MgO/Al$_2$O$_3$ (Mg: Al =3)	15:1	Soybean	MR	0.01	0.07	66	Hydrotalcites	Xie & Huang, 2006
Mgo/CaO (Mg:Ca =3.8)	12:1	Sunflower	60	0.04	0.92	92		Monica et al., 2008
MgO/Al$_2$O$_3$ (Mg:Al =5.8)	12:1	Sunflower	60	0.03	0.65	65		Monica et al., 2008
CaTiO$_3$	6:1	Rapeseed	60	0.01	0.08	79	Metal oxide catalysts	Kawashima et al., 2008
CaMnO$_3$	6:1	Rapeseed	60	0.01	0.09	92		Kawashima et al., 2008
Ca$_2$Fe$_2$O$_5$	6:1	Rapeseed	60	0.01	0.09	92		Kawashima et al., 2008
CaZrO$_3$	6:1	Rapeseed	60	0.01	0.09	88		Kawashima et al., 2008
CaCeO$_3$	6:1	Rapeseed	60	0.01	0.09	89		Kawashima et al., 2008
ETS-10	6:1	Soybean	100	0.01	0.31	92	Basic zeolites	Suppes et al., 2004
KOH/NaX	6:1	Soybean	60	0.01	0.03	82		Suppes et al., 2004
NaO/NaX	10:1	Soybean	65	0.01	0.03	28		Suppes et al., 2004
CsX	6:1	Soybean	60	0.001	0.01	7.3		Suppes et al., 2004
KI/NaX	15:1	Soybean	MR	0.01	0.02	12.9		Xie et al., 2006a
KI/ZnO	15:1	Soybean	MR	004	0.12	72.6	Alkali metal salts loaded on metal oxide	Xie et al., 2006a
CaO/ZnO	30:1	Palm kernel	60	0.01	0.31	93.5		Ngamcharussrivichai et al., 2008
KF/ZnO	15:1	Soybean	MR	0.03	0.19	80		Xie & Huang, 2006
KI/ZrO$_2$	15:1	Soybean	MR	0.04	0.13	78.2		Xie et al., 2006a

Maximum FAME yield (MFY) = FAME (g)/Oil (g). Catalyst (g), Maximum FAME productivity (MFP) = FAME (g)/Oil (g), t (h) FAME Yield (FY) = FAME (g)/ Oil (g), M/O = methanol–oil, reaction temperature (RT), methanol reflux temperature (MRT)

[a]n-hexane used as a co-solvent

[b]Reaction preformed in methanol supercritical conditions

[c]Reaction carried out under microwave irradiation, all reactions performed in a batch reactor

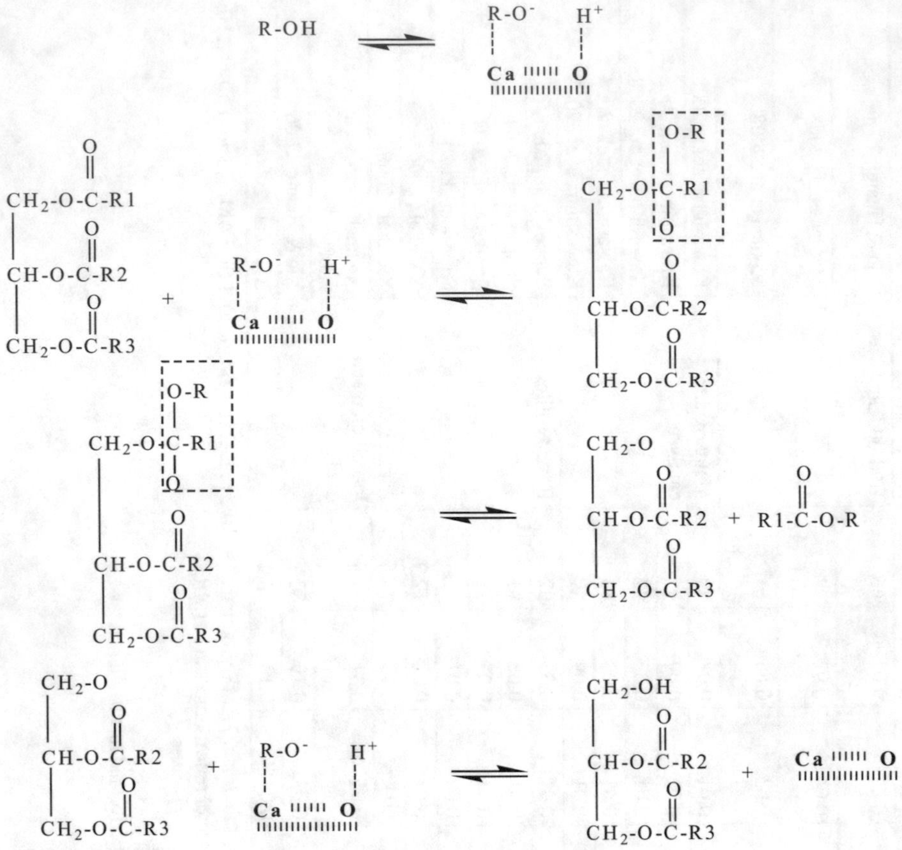

Fig. 2.7 Reaction route of transesterification of triglyceride with methanol using CaO. Source: Kouzu et al. (2008)

The calcium methoxide catalyst has been tested with 93 % of yield for the transesterification of oil. The dissociation of the methoxide gave rise to the catalytically active, strong basic CH_3O^- species that may increase the reaction rate. The reaction mechanism has been proposed by Tanabe, Misono, Ono, and Hattori (1989), as shown in Fig. 2.8. Besides the recycling experiment results showed, it has a long catalyst lifetime and can maintain activity even after being reused for 20 cycles (Liu, Piao, Wang, Zhu, & He, 2008c).

The SrO catalyst derived from thermal decomposition of $SrCO_3$ at 1200 °C was found to give high yield of FAME (Liu et al., 2007). Moreover, the long catalyst lifetime of SrO would maintain its activity even after being repeatedly used for ten cycles, which could be a commercially viable way to decrease the costs of production for industrial application in the transesterification of vegetable oils to biodiesel. Similarly, sodium silicate catalysts have also claimed to have 90 % of biodiesel yield at moderate temperatures (60–120 °C) (Portnoff et al., 2006), but there was no data on the catalyst's reusability.

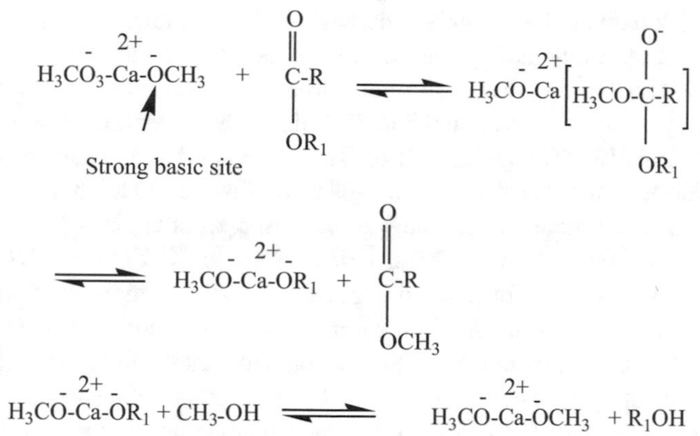

Fig. 2.8 Ester alcoholysis mechanism proposed on calcium methoxide catalysts. Source: Tanabe et al. (1989)

Hydrotalcites are Mg^{2+}/Al^{3+} layered double hydroxides with the general formula $Mg_{1-x}Al_x(OH)_2(CO_3)_{x/2}$ nH2O, and they are calcinated to give amorphous, porous metal oxides (PMOs) upon the evolution of H_2O and CO_2 (MacLeod et al., 2008). Using commercial hydrotalcite, MgO/Al_2O_3 (Mg:Al$=3$), soybean oil underwent up to a 66% conversion when refluxing methanol for the transesterification. In another study (Serio et al., 2006), temperatures above 180 °C were required to achieve yields higher than 90%. Nevertheless, other studies using MgAl at 60 °C have also shown conversions greater than 90%. Pure MgO exhibits weak base sites; therefore, the incorporation of small amounts of Al^{3+} cations to MgO drastically increased the generation of new surface Lewis acid–strong base pair site activities (Cosimo, Díez, Xu, Iglesi, & Apesteguı, 1998). This high performance indicates that the reaction is influenced by the base site as well as the surface area of the catalyst (Monica et al., 2008).

Kawashim et al., (2008) tested different kinds of metal oxides containing Ca, Ba, Mg, or La for the transesterification of oil at 60 °C with a 1:6 molar ratio of the oil to methanol for 10 h in a batch-type reactor. Positive results were obtained using $CaMnO_3$, $Ca_2Fe_2O_5$, $CaZrO_3$, and $CaCeO_3$, with the catalytic basicities ranging from 7.2 to 9.3 and showing 92, 92, 88, and 89% yields of the ester, respectively. In particular, a high durability of catalytic activity was found for the catalyst samples of $CaZrO_3$ and $CaO–CeO_2$, which were able to provide methyl ester yields greater than 80% for up to five and seven repeated uses, respectively. Unfortunately, these could not be investigated further, particularly with regard to the leaching of the catalyst during the reaction.

The investigation into the activity of zeolite in the transesterification reaction was reported by Suppes, Dasari, Doskocil, Mankidy, and Goff (2004). After 24 h, the conversion of soybean oil to methyl ester increased from 15.4 to 22.5% for potassium zeolite (KX) and 18.7% for cesium zeolite (CsX). The decrease in the conversion of CsX compared to that of KX was related to the degree of ion exchange for these solids. The large size of cesium cations limited the exchange capacity compared to that for the smaller potassium, which affected the basicity associated with the framework oxygen. However, an increase in the zeolite basicity could have also been achieved by impregnating the porous solids with cesium oxide (Hathaway & Davis, 1989) and thus forming additional framework basic sites. Interestingly, a similar phenomenon was observed in the case of the ETS-10 zeolites (Engelhard titanosilicate structure-10) (Kuznicki, 1989). The Engelhard titanosilicate structure-10 (ETS-10) formed more basic sites, and this advantage led to a higher conversion to the methyl esters compared to the NaX zeolite (Suppes et al., 2004).

Alkali metal salts loaded on metal oxides, such as KI/ZnO and KI/ZrO$_2$, have been cited as alternatives for performing the transesterification reaction Xie et al., 2006a). It was found that the KI/ZrO$_2$ catalyst was more active compared to the KI/ZnO in the transesterification of vegetable oil with methanol. It is likely that the activity of the catalysts is strongly affected by the strength of basic sites of the catalysts (Xie et al., 2006a). Therefore, in this case, the increase in the base's strength is thought to be from the formation of O$^-$ centers that arise by the substitution of M$^+$ ions into the alkaline earth oxide lattice (Baronetti, Padro, Scelza, & Castro, 1993). Favorably, a higher transesterification activity (>94 %) was obtained using a proper combination of mixed oxides (CaO/ZnO) derived from the corresponding mixed metal carbonate precursors (Ngamcharussrivichai et al., 2008).

Among the basic heterogeneous catalysts, the Ca(OH)$_2$ and Ca(OCH)$_2$ catalysts gave a high conversion, which was explained by the presence of stronger base sites in the catalyst. The formation of basic sites on a supported catalyst has several explanations: (a) The solid-state reaction between the guest compound and the surface of the support in the activation process is favorable for the catalyst's basicity. In other words, the metal ion of catalyst could insert in the vacant sites of the support, accelerating the dissociative dispersion and decomposition of the catalyst to form the basic sites during the activation process (Xie et al., 2006a). (b) The more catalyst that is loaded on the support, the fewer free vacancies available, which results in the surface enrichment of the metal species. This surface enrichment is probably the location of the active sites for base-catalyzed reactions (Xie et al., 2006a; 2006b). (c) When the amount of metal ion loaded on support was above the saturation uptake, it could not disperse well and, for this reason, only a part of the loaded catalyst could decompose. As a result, the number of basic sites, together with the activities of the catalysts, would decrease (Meher et al., 2006; Xie et al., 2006a). Thus, the basic sites would be proportional to the decomposed catalyst, instead of proportional to the loaded catalyst. As a general conclusion, it is possible to say that a large variety of solid, base catalysts are now available.

However, it is not possible, with any of the catalysts presented in this review, to possess simultaneously a strong, basic, high surface area, inexpensive catalyst production, and catalyst stability. Certainly, a compromise needs to be reached in each case.

2.2.11.3   Solid Acid Catalysts

The acidity of a material is measured relative to a base in the acid–base interaction. In the case of Brönsted acidity, the solid acid is able to donate or at least partially transfer a proton, which becomes associated with surface anions. By the Lewis definition for an acid, a solid acid must be able to accept an electron pair. Thus, when the acid surface reacts with a Lewis base molecule, a coordinate bond is formed. Heterogeneous, Lewis acid catalysts favor the formation of electrophilic species, which determine the rate of desorption followed by the rate of the transesterification reaction (Serio et al., 2005). Here, the strong, Lewis acidic sites favor a low desorption rate, resulting in a slow transesterification reaction (Martyanov & Sayari, 2008). The influences of temperature, catalyst quantity and reaction time using solid, acid catalysts on FAME yields are summarized in Table 2.6.

Almeida et al. (2008) tested the yield of biodiesel using different ratios of TiO_2–SO_4 and the following order of reactivity was obtained: TiO_2–$SO_4(5:1) > TiO_2$–$SO_4(10:1) > TiO_2$–SO_4 (20:1). The low activity of the 20:1 TiO_2–SO_4 ratio must be related to the low Brönsted acid amount on its framework, which is probably due the insufficient amount of sulfuric acid used in its preparation. However, the enhancement of Brönsted acid sites in a catalyst was reported by He, Baoxiang, Dezheng, and Jinfu (2007). He et al. observed that TiO_2 modified with sulfate was found to be an active catalyst for the transesterification of cottonseed oil with methanol, due to the creation of new Brönsted acid sites (Corma, 1995; Yamaguchi, 1990), as shown in Fig. 2.9. It was also reported by Kiss, Omota, Dimian, and Rothenberg (2006) that higher sulfur content corresponds to higher acidity of the catalyst and consequently higher catalytic activity. Overall, the results are in good agreement with the literature (Ardizzone, Bianchi, Ragaini, & Vercelli, 1999; Matsuda & Okuhara, 1998).

Jitputti et al. (2006) highlighted the use of acid catalysts ZrO_2 and SO_4^{2-}/SnO_2 in the transesterification of palm oil and coconut oil with methanol. The SO_4^{2-}/SnO_2 super-acid, solid catalyst gave a higher yield of methyl esters when compared to the ZrO_2 catalyst. Specifically, the SO_4/ZrO_2 catalyst showed superior initial catalytic activity, but it rapidly deactivated, possibly due to the sulfate leaching that caused deactivation of the active sites (Garcia et al., 2008).

Tungstated zirconia (WO_3/ZrO_2) is another promising solid, acid catalyst for the production of biodiesel fuels from soybean oil because of its high performance (over 95 % conversion) in the transesterification reaction (Furuta et al., 2004). In particular, the tetragonal phase of the ZrO_2 displayed better performance, reaching a maximum conversion of 95 %; whereas, the monoclinic phase of ZrO_2 obtained less than 5.5 times the conversion (Ramu et al., 2004). The reason this discrepancy occurred is that the tetragonal phase of ZrO_2 has a higher content of non-bridging surface hydroxyl groups than the monoclinic phase (Ward & Ko, 1995), which seems to be a crucial factor for producing more active catalysts.

SnO catalysts showed a higher conversion (95 %) in the transesterification of vegetable oil with methanol compared with $SnCl_2$ (Abreu, Alves, Macedo, Zara, & Suarez, 2005; Vicente, Coteron, Martinez, & Aracil, 1998). It was found that the SnO catalyst retained high proportion of its original activity over time. Additionally, a Zeolite Y catalyst also has the possibility to be recycled without any loss of its catalytic activity (Brito et al., 2007). On the other hand, deactivation tests have yet to be performed with longer reactions in a higher scale reactor to verify this fact.

Table 2.6 Summarization of biodiesel synthesis using acidic heterogeneous catalyst

Catalyst	M/O	Oil	RT (°C)	MFY	MFP	FY (%)	P (bar)	CT (°C)	References
WZA	40:1	Soybean	250	0.22	0.044	88	1.013	800	Furuta, Matsuhashi, & Arata, 2006
TiO_2/ZrO_2	40:1	Soybean	250	0.21	0.042	84	1.013	800	Furuta et al., 2006
WO_3/ZrO_2^1	40:1	Soybean	>250	0.0030	0.225	90	1.013	800	Furuta et al., 2004
SO_4/ZrO_2^1	40:1	Soybean	300	0.0027	0.20	80	1.013	675	Furuta et al., 2004
SO_4/ZrO_2	20:1	Soybean	150	0.0991	0.4335	86.7	-	600	Garcia et al., 2008
ZrO_2/SO_4	12:1	Cottonseed	230	0.04	0.10	80	-	550	He, Baoxiang, et al., 2007
SnO^{2-}/ZrO_2	6:1	Coconut	200	0.0426	0.2157	86.3	50	500	Jitputti et al., 2006
SO_4^{2-}/SnO_2	6:1	Coconut	200	0.039	0.2014	80.6	50	500	Jitputti et al., 2006
MgO/SBA-15	-	Vegetable	220	0.96	0.192	96	-	550	Li, Xu, & Rudolph, 2009
MgCoAl-LDH	[a]16:1	Canola	200	0.543	0.2083	96	25.331	600	Li et al., 2009
Al-MCM-41	60:1	Palmitic	180	0.4038	0.3105	60	-	480	Jr. et al., 2009
Nano-MgO	36:1	Soybean	170	0.0373	0.539	98	240	-	Wang & Yang, 2007
s-MWCNTs	18.2:1	Cottonseed	260	0.5187	0.2999	89.93	-	700	Shu et al., 2009
KNO_3 loading flayish	15:1	Sunflower	170	0.0199	0.1099	87.5	-	500	Kotwal, Niphadkar, Deshpande, Bokade, & Joshi, 2009
Montmorillonite KSF	8:1	Palm	190	0.030	0.265	79.6	-	-	Kansedo, Lee, & Bhatia, 2009
$ZnO-La_2O_3$	3:1	Palm oil	200	0.32	0.32	96	-	450	Yan, Salley, & Ng, 2009
ZS/Si	18:1	WC	200	0.27	0.081	81	-	110	Jacobson, Gopinath, Meher, & Dalai, 2008

VOP	Soybean	27:1	180	0.08	0.8	80	-	300	Serio et al., 2007
Fe-Zn-DPEG-4000	Sunflower	15:1	250	0.0356	0.241	96.5	-	180	Sreeprasanth et al., 2006
BMZ	Soybean	32.7:1	450	0.0025	0.463	92.6	172.369	444	McNeff et al., 2008
V_2O_5/TiO_2	Palm	45:1	120	0.1434	0.0742	74	-	460	Ratanawilai, Suppalukpanya, & Tongurai, 2005
Pb_3O_4	Soybean	7:1	225	0.41	0.41	82	high	-	Singh & Fernando, 2008
MgO	Soybean	55:1	130	0.0186	0.1170	82	-	-	Antunes et al., 2008
MgO	Soybean	0.44:1	>180	0.92	0.92	92	-	500	Serio et al., 2006
MgO	Soybean	39.6:1	300	0.08	5.69	91	-	-	Tateno & Sasaki, 2004

Sulfonated multi-walled carbon nanotubes (s-MWCNTs), zinc stearate (ZS), vanadyl phosphate(VOP), base modified zirconia (BMZ), zinc hydroxide nitrate (ZHN), polyethylene glycol (PEG), zinc sulfate (ZS), layered double hydroxides (LDH), amorphous zirconium tungstate (WZA), Santa Barbara Amorphous type material (SBA-15), Mobil Composition of Matter No. 41 (MCM-41)

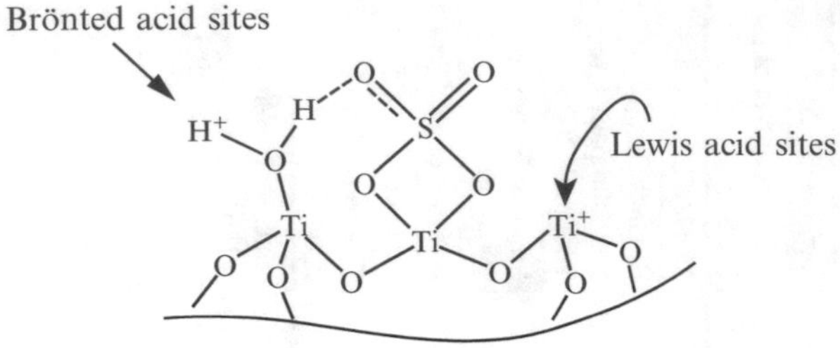

Fig. 2.9 Structure of
Brønsted acid and Lewis
acid sites on sulfated
titania. Source: Corma
(1995) and Yamaguchi
(1990)

2.2.11.4   Effect of Catalyst Hydrophilicity and Hydrophobicity

The hydrophilic/hydrophobic character of the catalyst surface had received a great deal of attention with regard to the mesoporous surface reactivity of catalysts. It is obvious that the catalyst surface should posses some hydrophobic character to promote the preferential adsorption of oily hydrophobic species on the catalyst surface and to avoid deactivation of catalytic sites by strong adsorption of polar by-products like glycerol or water (Melero, Grieken, & Morales, 2006). Moreover, most solid, acid catalysts are absorbed by water or glycerol in reactions where access to fatty acid molecules is hindered (Lotero et al., 2005). Thus, the hydrophobicity of the acid sites is an important challenge in order to reduce poisoning by polar molecules.

A strategy for controlling the hydrophobicity of zeolite was to use the Si/Al ratio method that has been developed by Kiss et al. (2006). This attempt was substantially improved by the creation of reaction pockets inside the hydrophobic environment by adjusting Si–Al ratio so the fatty acid molecules could be absorbed and the polar molecules would avoid being absorbed (Kiss et al., 2006). Another approach recently introduced utilized the formation of a hydrophobic catalyst through the sulfonation of incompletely carbonized starch (Lou, Zong, & Duan, 2008). Lou et al. mentioned that the catalyst led to the successful conversion of waste cooking oils containing high fatty acid esters to biodiesel.

Similarly, an approach for the preparation of fatty acid alkyl esters (biodiesel) from waste cooking oils and nonedible oils was proposed by Sreeprasanth, Srivastava, Srinivas, and Ratnasamy (2006). In this case, the enhanced catalytic performance in the transesterification reaction was explained by the improved accessibility of the active sites (Zn^{2+}) and the hydrophobic organic groups into the mesoporous silica. Overall, numerous strategies for preparation of sulfonic modified mesoporous hydrophobic groups have been developed, as shown in Table 2.7. Mesoporous catalysts containing methyl, propyl and silane modified SO_3H MCM-41 have been reported to be efficient catalysts (95 % conversion) in the esterification of glycerol with fatty acids (Dıaz et al., 2003). The direct co-condensation synthesis, in which the mesostructure and functional group are simultaneously introduced, appears to be a desirable route for incorporating the functional group because it increases the concentration of sulfonic groups in the mesoporous silica relative to the post formation grafting (Dıaz et al., 2003).

In order to modify the hydrophobicity of the microenvironment surrounding the sites of perfluorosulfonic-acid, Bols and Skrydstrup (1995) incorporated a propyl functional group (via incorporation of an appropriate triethoxysilane) during the

Table 2.7 Sulfonic modified mesoporous hydrophobic groups

Catalyst	Process	References
Silane modified SO₃H MCM-41	Co-condensation of organic functionalities with sulfonic groups on the silica surface	Diaz et al., 2003
Methayl modified SO₃H MCM-41	In situ oxidation	Diaz et al., 2001
Perfluoroalkyl sulfonic-acid modified MCM-41	Grafting of the precursor (1,2,2-trifluoro-2-hydroxy-1 trifluoromethylethane sulfonic acid sultone) over the silica surface	Bols & Skrydstrup, 1995
Alkanesulfonic-functionalized PMO	One-step direct synthesis with in situ oxidation	Melde et al., 1999
Ar-SO₃H SBA-15	Acylation processes	Mbaraka et al., 2003
Ar-SO₃H MCM-41	Grafting of phenyl groups to the silica surface and subsequent sulfonation with chlorosulfuric acid	Lindlar et al., 2001

Mobile Crystalline Material (MCM-41), periodic mesoporous organosilica (PMO), Santa Barbara Amorphous type material (SBA-15)

synthesis. An approach to improve the performance of alkanesulfonic-functionalized Periodic Mesoporous Organosilica (PMO) is to create a reaction environment that will continuously exclude water from the mesopores by incorporating organic groups and manipulating the synthesis conditions (Melde, Holland, Blandford, & Stein, 1999). The combination of both functionalities, acidic (–SO₃H) and hydrophobic (-R-), in the PMO materials results in interesting surface properties that could enhance the diffusion of reactants and products in the acid catalyzed reaction (Melero et al., 2006), as shown in Fig. 2.10.

The strength of the acid sites can be increased by incorporating functionalized groups, either by grafting on the silica surface or by a post synthesis route process (Mbaraka, Radu, Lin, & Shanks, 2003). The presence of an arene sulfonic acid in the MCM-41 structure significantly increases the acid strength of the material, which becomes comparable to that of a pure MCM-41 catalyst (Lindlar et al., 2001). In this approach, a swelling agent was used as a binder for assembling the ordered mesoporous structure. However, the swelling agent is soluble in the hydrophobic part of the catalyst, leading to the formation of a less organized mesoporous structure. Indeed, trimethylammonium solution (TMA⁺) could be used to reduce swelling in the hydrophobic part of the MCM-41 catalyst (Passerini, Coustier, Giorgetti, & Smyrl, 1999).

2.2.11.5 Effect of Calcination Temperature of Catalyst

It has been found that the catalyst activity can be restored and the catalyst performance improved by calcination. The calcination temperature is crucial for the generation of catalytic activity. Moreover, there have been attempts to calcinate the catalysts at various temperatures since the calcination process has been found to

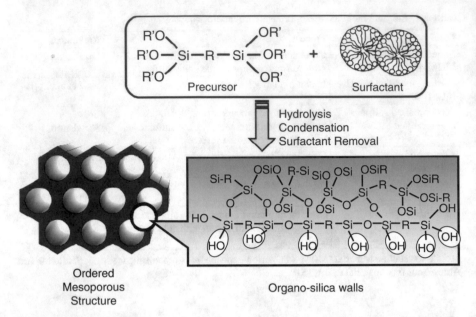

Fig. 2.10 Strategy of synthesis for the preparation of PMO's materials. Source: Melero et al. (2006)

affect the structural and catalytic properties of catalysts. Benjapornkulaphong et al. evaluated the effect of the calcination temperature for various Al2O3-supported metal oxides on the methyl ester (ME) content of the biodiesels formed from the transesterification of palm kernel oil, as shown in Fig. 2.11.

It is clearly seen that $Ca(NO_3)_2/Al_2O_3$ exhibited a much higher conversion (94 %) at 450 °C compared with other catalysts. Nevertheless, a substantial decrease in catalytic activity was explained by a considerable increase of the calcination temperature (>550 °C). A similar observation has been made by Wan et al. who attempted to relate the calcination temperature of the catalysts to the yield of biodiesel, and they have suggested that the decrease in catalytic activity was perhaps due to the adsorption of organic materials, which led to the carbonization on the catalyst surface at a high temperature.

Yan, Lu, and Liang (2008) reported that high yield of biodiesel (92 %) can be obtained in the transesterification of rapeseed oil at 64.5 °C using calcined CaO/MgO, even if the CaO/MgO catalyst was contaminated by H_2O, O_2, CO_2, and other gaseous substances contained in air during storage, which could reduce the catalytic activity in the transesterification reaction. Additionally, the completely deactivated CaO catalyst derived from an eggshell could be regenerated by a simple calcination at 600 °C (Wei, Xu, & Li, 2009). It is particular interesting to note that this advantage was maintained even after being used for 13 times.

Xie and Huang, (2006) and Xie et al., (2006a; 2006b) extensively studied the transesterification of vegetable oil over supported catalysts under different calcina-

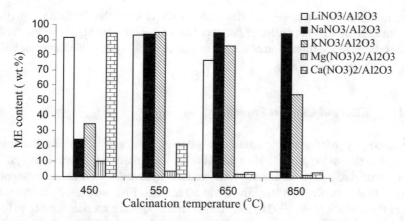

Fig. 2.11 Percentage of FAME yields obtained by different calcination temperature in the transesterification reaction of palm kernel oil. (methanol–oil molar ratio, 65; catalyst amount,10 wt% based on oil weight; temperature, 60 °C; time 3 h). Source: Benjapornkulaphong et al. (2009)

tion temperatures. The authors reported that optimization of the calcination temperature in the preparation of catalysts is essential to achieve high catalytic performance. Catalysts calcined at 500–600 °C, such as KI/Al_2O_3 and KNO_3/Al_2O_3, showed 87 % of biodiesel yield. The catalysts were active in the transesterification reaction due to active species (K_2O) formed after calcination. However, the recyclability of the catalysts was not fully investigated, particularly the stability of the catalyst in the reaction environment at high temperatures. Higher activity active species, such as Li_2O compared with the K_2O species, were explained. The presence of stronger basic sites associated with the Li/ZnO gave Li_2O greater activity (Shumaker et al., 2008; Xie et al., 2007). However, its reusability was limited due to catalyst deactivation.

Kawashim et al., (2008) performed a detail comparison study of the calcium-containing catalysts: $CaMnO_3$, $Ca_2Fe_2O_5$, $CaZrO_3$, and $CaO–CeO_2$ by performing the transesterifications at 60 °C with a 6:1 molar ratio of methanol to oil for 10 h. In this case, the enhanced catalytic performance was explained by the improved accessibility to the active calcium methoxide species formed after calcination at 900 °C. The CaO/ZnO and KF/ZnO catalysts were reported to be inactive as the calcination temperature increased beyond 800 °C (Ngamcharussrivichai et al., 2008; Xie & Huang, 2006). This inactivity might be due to the sublimation or penetration of the active species of the CaO/ZnO and KF/ZnO catalysts into the subsurface of the support at 800 °C, which in turn reduced the catalytic activity (Lukic et al., 2009; Xie & Huang, 2006).

The activity of the WO_3/ZrO_2 catalyst for the esterification of palmitic acid with methanol was found to be linked with the calcination temperature (Ramu et al., 2004). When the calcination temperature increased from 500 to 800 °C, a significant increase in catalytic activity was attributed to the phase transformation of ZrO_2

(from monoclinic to tetragonal). However, in the case of the ZrO_2/SO_4 catalyst, a substantial increase in the yield of fatty acid methyl esters (FAME) over the catalyst can be related to the formation of new Brönsted acid cites after calcination at 550 °C (He, Baoxiang, et al., 2007).

2.2.11.6   Effect of Catalyst Porosity

The discovery of the mesoporous catalyst has generated great interest among researchers due to the prospective application of these materials as solid catalyst supports, particularly for their liquid-phase reaction systems. The ordered mesoporous material with tunable pore sizes from 2 to 50 nm was first reported by Mobil scientists in 1992 (Beck et al., 1992). However, the mesoporous materials have been found to have relatively low catalytic activity and hydrothermal stability when compared with the conventional zeolites, which severely hinder their practical applications in catalytic reactions for the petroleum industry (Xiao, 2004). In other words, there is a challenge to prepare a new type of catalyst that should benefit from the advantages of the mesoporous materials.

There have been a number of successful examples for preparing thermally stable catalysts in the literature (Ryoo, Jun, Kim, & Jim, 1997). Good catalytic performance of the sulfonated, ordered mesoporous carbon (OMC) was attributed to the combination of strong acidic sites located inside the mesopores and the high stability of the OMC under the reaction conditions (Liu, Wang, Zhao, & Feng, 2008a). The OMC was synthesized by the covalent attachment of sulfonic acid containing aryl radicals on the surface of mesoporous carbon, followed by drying at 100 °C. In this respect, Di et al. (2003) performed a detailed comparison study of the ordered mesoporous aluminosilicates (MAS) and Al-MCM-41 catalysts. High catalytic activity stability was explained by the presence of the hexagonal mesoporous structure associated with the MAS and Al-MCM-41 catalysts.

Due to the relatively small size of the pore for the zeolite and zeotype materials (the maximum pore size is typically 1.2 nm), much effort has been devoted to the development of new materials with pores larger than 1.5 nm (Corma, Diaz-Cabanas, Jorda, Martinez, & Moliner, 2006). Mbaraka and Shanks (2006) have suggested that the presence of such material is important due to its three dimensional mesoporous channeling structure, which is more resistant to pore blockage and allows diffusion of the triglyceride molecules. As an example, the transesterification of soybean oil was investigated using immobilized alkylguanidines (1,5,7-triazabicyclodec-5-ene and 1,2,3-tricyclohexylguanidine) onto polystyrene, MCM-41 and zeolite Y (Sercheli, Vargas, & Schuchardt, 1999). The author showed that the MCM-41 catalyst exhibited a lower conversion compared with polystyrene. This was attributed to the presence of a small pore size (>1 nm) that limited the diffusion of the oil molecules into the catalytic sites situated inside the porous channel and thereby reduced the conversion of biodiesel. However, the mesoporous structure can be controlled by a sophisticated choice of templates (surfactants), and the pores of the MCM-41 can be increased by adding auxiliary organic chemicals (e.g., mesitylene) or changing the reaction parameters (e.g., temperature and compositions) (Beck et al., 1992).

Mesoporous calcium methoxide, H_3PO_4/Al_2O_3 catalysts have been used in the transesterification reaction with alcohol, and it obtained biodiesel with greater than a 96% yield (Araujoa, Scofielda, Pasturaa, & Gonzalezb, 2006). The high yield for biodiesel could be demonstrated by the surface area of the catalyst being occupied by pores of relatively large size (between 100 and 1400 nm). For smaller pored materials, the accumulation of methanol and water around the hydrophilic sites hinder the access of free fatty acid ester (FFA) molecules to catalyst, giving rise to a diffusion controlled reaction. For instance, the higher activity of calcined hydrotalcite (CHT) with a Al/(Al+Mg) ratio of 0.2 compared to MgO was due to the larger pores (>20 Å) in the CHT compared to the MgO (<10 Å). The larger pores rendered the active sites so they were accessible by the bulky triglyceride molecules (Serio et al., 2006). Thus, the porous, solid catalyst offered advantages over the traditional solid catalysts because of their improved mass-transfer characteristics for both reactants and products in the catalytic reaction.

2.2.11.7 Effect of Catalyst Surface Area

A great deal of effort has been devoted in recent years to obtaining catalysts with high surface areas, owing to its wide application potential in the field of petrochemical industries (Araujoa et al., 2006; Buchmeiser, 2001; Kirkland, Truszkowski, Dilks, & Engel, 2000). In particular, it has been observed by Lopez, Suwannakarn, Bruce, and Goodwin (2007) that the surface area of the catalysts was influenced by the calcination temperature. For instance, a 325-m^2/g surface area for tungstated zirconia was prepared by dehydration of a WZ precursor at 120 °C and then pro treatment for 3 h under an air atmosphere in a furnace at 800 °C. However, after calcination at 900 °C, the surface area was only about 20% WZ (58-m^2/g). The loss of surface area was explained by the loss of the tetragonal phase for the ZrO_2 structure at 900 °C (Lopez et al., 2007). Overall, the results are in good agreement with the previously reported BET measurements for tungstated zirconia (Boyse & Ko, 1999; Garcia et al., 2008).

He, Baoxiang, et al. (2007) studied the influence of surface area for TiO_2, ZrO_2, TiO_2-SO_4^{2-} and ZrO_2-SO_4^{2-} catalysts in the transesterification of cottonseed oil with methanol at 230 °C. It was shown that the specific surface areas of TiO_2 and ZrO_2 were increased by incorporation of sulfate to both TiO_2 and ZrO_2 structure. Consequently, the large accessible surface area improved the efficiency of the transesterification.

Mixed oxides of Ca and Zn also proved to be a promising catalyst in the transesterification of palm kennel oil (Ngamcharussrivichai et al., 2008). It was shown that CaO/ZnO derived from the corresponding mixed metal carbonate precursor had 94% of biodiesel yield. It was reported by Ngamcharussrivichai et al. (2008) that the high surface area of the metal-carbonate could be responsible for high biodiesel yield

(94 %). Therefore, a catalyst with high surface area could make the reactions proceed at rates high enough to permit their commercial exploitation on a large scale.

2.2.11.8 Effect of Catalyst Reusability

Reusability is one of the most important properties of a solid catalyst. The reuse of the solid catalyst is governed by their deactivation, poisoning, and the extent of leaching in the reaction medium (Lee, Park, & Lee, 2009). Leaching affects the industrial application as extensive leaching may threaten the reusability and the environmental sustainability of catalyst (Granados, Alonso, Sadaba, & Ocon, 2009). Moreover, the degree of leaching directly affects the number of runs that the catalyst can be reutilized when operating in batch-wise mode or the time of operation when working in a continuous process. Therefore, the development of reusable solid catalysts having high activity for the liquid phase reactions is of great practical importance.

The deactivation tests usually take the form of repeating the reaction cycle several times and measuring the yield of biodiesel in the interval between each cycle. If the deactivation of the catalyst is unavoidable, a method for regenerating its initial activity was suggested in most cases (Lee, Park, et al., 2009). Ma et al. (2008) have investigated the catalytic stability of heterogeneous base catalyst, γ-Al_2O_3, loaded with K and KOH in the transesterification of rapeseed oil with methanol. The catalyst after regeneration exhibited a lower conversion than before (84.5 vs. 46.92 %). The substantial decrease in catalytic activity was explained by a considerable leaching of the active species into the methanol phase, as found by Alonso, Mariscal, Tost, Poves, and Granados (2007). Similarly, the TiO_2/SO_4 catalyst has been used in the transesterification reaction of vegetable oil, and the deactivation of the catalyst was attributed to the leaching of active sulfate species (Almeida et al., 2008). However, sulfate species are sensitive to water vapor (Corma, 1997). For instance, the SO_3 moieties on the surface gave rise to sulfate species, such as SO_4^{2-}, HSO_4^-, and H_2SO_4. In a liquid phase system, these sulfate species form and easily leach out from the catalysts. In any case, it seems quite clear that the presence of sulfate is a problem or at least a limitation for the practical use of this type of catalyst.

Mootabadi, Salamatinia, Bhatia, and Abdullah (2010) investigated the ultrasonic-assisted transesterification of palm oil in the presence of alkaline earth metal oxide catalysts (CaO, SrO and BaO). It was found that BaO had the highest residual elements detected in biodiesel with nearly 14 wt% of the catalyst leached into the biodiesel layer after the reaction. Meanwhile, CaO had the minimum solubility with only 0.04 % weight loss. This result indicated that BaO had the highest solubility in biodiesel and appreciable leaching occurred. MacLeod et al. (2008) transesterified rapeseed oil under the following conditions: a 6:1 methanol–oil molar ratio, a $LiNO_3$/CaO, $NaNO_3$/CaO, or KNO_3/CaO catalyst and 60 °C. A yield of >90 % was observed after 3 h for all four types of catalysts. They have evaluated the stability of catalysts for application to biodiesel production, concluding that metal leaching from the catalyst was detected, and this resulted in some homogeneous activity. The authors pointed out that this drawback would need to be resolved before scaling-up biodiesel production. Similar results have been reported in the literature in reference to leaching during their transesterification reactions (Granados et al., 2007). Therefore, if any support of liquid phase processing is used, leaching of the supported catalyst can be a serious problem.

Because the deactivation of the catalysts is the main hurdle to the heterogeneous catalyzed process, several attempts have been made to develop a stable and effective catalyst for biodiesel production (Feng et al., 2010; Serio et al., 2008; Zong, Duan, Lou, Smith, & Wu, 2007). For instance, cation exchange resins (NKC-9, 001×7 and D61) were tried by Feng et al. (2010) and found to be effective in esterification of high acid value (13.7 mg KOH/g) feedstock of waste cooking oil (WCO) origin. NKC-9 had high water-adsorbing capacity favoring its role in effective esterification. A high average pore diameter of NKC-9 was helpful for reactants to access the active sites of the resin resulting in greater than 90 % conversion. The reaction conditions were 6:1 (alcohol to oil) molar ratio, 24 wt% of the catalyst at 64 °C for 4 h of reaction time. The catalyst NKC-9 further reused up to 10 runs. The activity of the catalyst in subsequent reuse did not deteriorate, but rather it was enhanced. This has been attributed to the breakdown of the resin particles by mechanical agitation, which increased the surface area of the resin. Aft er ten runs, there was loss of the catalyst during separation which ultimately decreased the free fatty acid (FFA) conversion, so new resin was added. Similarly, the sugar catalyst prepared by an incomplete carbonization process was found to be stable even after more than 50 cycles of successive use for biodiesel production from waste oil (Zong et al., 2007). Another approach to reduce the catalyst leaching is to use $VOPO_4.2H_2O$ in the transesterification process; the catalytic activity is restored by calcination in air. However, the authors also claimed that the high activity of $VOPO_4.2H_2O$ catalyst is related to the surface vanadium V^{5+} species, which can reduce to V^{3+} species after the reaction, which is associated with the deactivation of the catalysts (Serio et al., 2008). In conclusion, the stability of catalysts will need to be improved significantly before they are suitable for industrial use.

2.3   Conclusion

It may not be possible with any one of the catalysts presented in this review to possess simultaneously a strong acid/base, high surface area, porosity, and inexpensive catalyst production. Certainly, one needs to strike a balance by considering all the process parameters in each case. The extensive effort on the development of heterogeneous catalysts for biodiesel production, as summarized has led to an enhanced understanding on the chemical and physical properties of a catalyst that play a vital role on the biodiesel yield. Basically, base catalysis is a better choice than acid catalysis in terms of the reaction rate and biodiesel productivity. However, the adverse effect on the activity by water and FFA must be overcome before any developed catalyst can claim to possess robust activity for base-catalyzed biodiesel synthesis. Though, the high yield of biodiesel has been produced using solid base or acid catalyst, the catalysts are in the form of powders with diameter ranging from nanometer to micrometer. From the practical point of view, handling of small particles could be difficult due to the formation of pulverulent materials. Utilization of powders in catalytic reactions renders their recovery and purification, and energy intensive ultracentrifugation is needed for the subsequent separation operation. In addition, the active phase of the powdered catalyst may not be uniformly distributed on the support but rather form localized aggregates leading to low contact of active surface in the catalyst. Thus, the efficiency of the catalyst and its feasibility at industrial scale might be reduced. Based on the above discussions, developing a new solid the design of a catalyst at a millimetric seems to be an appropriate solution to overcome problems associated with the traditional catalysts. The mechanical strength and shape of the catalyst is the key issue for the millimetric heterogeneous catalysts. As previously stated, supported catalyst in spherical form can offer shape-dependent advantages such as minimizing the abrasion of catalyst in the reaction environmentIt is further highlighted that high mechanical strength is crucial for long-term stability of catalyst. Therefore, easy handling, separation, and reusability are the main strengths which could respond to select the spherical millimetric catalyst.

Chapter 3
Materials and Methods

3.1 Introduction

This chapter describes the materials and methods applied throughout this thesis, giving a description of the procedures that were used to prepare the catalyst particle. Besides, catalysts characterization technique and mathematical equations were also included into this chapter. The summary is indicated in the following flowchart (Fig. 3.1). Different materials used are provided in the text.

3.2 Determination of Rheological Properties of the Boehmite Sols

3.2.1 Preparation of Boehmite Suspension

Boehmite (AlOOH) powder was provided by BASF chemical company (G-250, medium density pseudoboehmite alumina, BASF Catalysts LLC in USA, purity >99 %). The properties of AlOOH are summarized in Table 3.1. The hydrochloric acid used was of analytical grade. The boehmite powder was mixed with deionized water at different concentrations ranging from 50 to 300 g/L. The suspensions were dispersed by sonication for 2 min. For sonication, a microprocessor based and programmable ultrasound processor was used (Autotune series, model 750 Watt). This processor had a frequency of 20 kHz with maximum power output of 750 W. In addition, the processor had the facilities of amplitude controller, which was set at 70 % during experiments. Figure 3.2 shows the setup of sonication apparatus used in this study. After ultrasonic treatment, homogeneous and lightly opalescent boehmite sol was obtained. Subsequently, the suspensions were stirred by a mechanical stirrer for a period of 20 min at a speed of 150 rpm. The suspensions were then left unperturbed for an additional 30 min for removal of air bubbles and better homogeneity,

© Springer International Publishing Switzerland 2017
A. Islam, P. Ravindra, *Biodiesel Production with Green Technologies*,
DOI 10.1007/978-3-319-45273-9_3

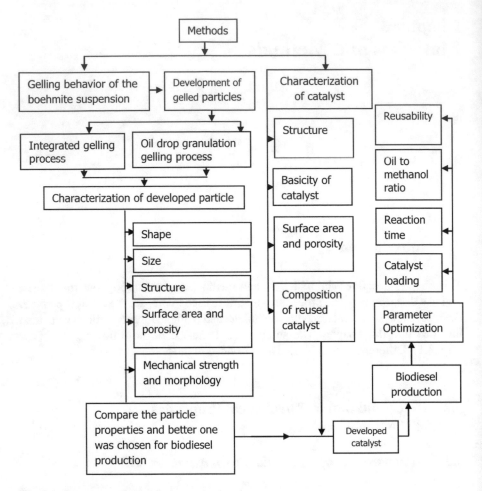

Fig. 3.1 Summary of materials and methods

which is essential prior to the rheological measurements. The suspensions were further acidified to different pH (1.0–7.6) using pH meter (Cyber Scan pH 1500, Eutech Instruments Pte Ltd, Singapore) at 25 °C by the addition of concentrated hydrochloric acid. Then suspensions were evaluated with respect to their rheology, by means of viscosity measurements.

3.2.2 Particle Size Measurement

Dynamic laser diffraction particle size distribution Analyzer (HORIBA; LA-300) was used to measure the size distributions of the boehmite particles. The suspension was prepared by dispersing the powders in distilled water at a constant temperature of 25 °C. The suspension was then treated to ultrasonic vibrations for a few minutes

Table 3.1 The chemical compositions of AlOOH[a]

AlOOH (Wt.%)	Impurities (wt.%)				
	SiO_2	Fe_2O_3	Na_2O	CaO	SO_4
>99.6	0.02	0.01	0.03	0.03	0.30

[a]Information provided by the manufacturer

Fig. 3.2 Sonication apparatus. *1.* Microprocessor based control unit of ultrasonic; *2.* Ultrasound horn; *3.* Sample used for sonication

to declump the clusters. This equipment is operated through the LA-300 software package that performs all the functions, such as making measurements, and storing and retrieving data. The LA-300 software automatically calculates the particle size distribution.

3.2.3 Measurement of Rheological Properties of Boehmite Suspensions

The rheological properties of the boehmite suspensions were determined using a rheometer (Programmable DV III+ Rheometer, Brookfield Engineering Laboratories, Inc., USA). The appropriate disc spindles HA/HB-2, HA/HB-3 were used for viscosity measurement at zero shear rate, whereas the spindles SC4-18 and SC4-28 were used for viscosity measurements at different shear rates. The rheometer was checked with the calibration fluids each time before use. The rheometer was connected to a water

Fig. 3.3 Rheological properties measurement apparatus. *1*. Viscometer (Brookfield Engineering Laboratories, Inc., USA. Model: Programmable DV III+); *2*. Water bath with circulation pump (Thermo/Haake DC30-K10)

bath with circulating pump (Model: Thermo/Haake DC30-K10) to control the suspension temperature from 25 to 75 °C as shown in Fig. 3.3. The flow curves (shear stress vs. shear rate, viscosity vs. shear rate) were determined at shear rates in the range between 10 and 300 s^{-1}. Measurements were repeated three times for each sample.

3.2.4   Determination of Consistency Index (k) and Flow Behavior Index (n)

The influence of different pHs on the viscosity of suspensions were then investigated by using the Oswald-De Waele model given by Eq. (3.1), also known as the power-law model (Rao & Anantheswaran, 1982).

$$T = k(\gamma)^n \tag{3.1}$$

where τ is the shear stress in dyne per cm^2, γ is the shear rate per second (s^{-1}), k is the consistency index (Pa.s), and n is the flow behavior index. A linear regression was carried out on the linearized form of Eq. (3.1) using the experimental shear rate-shear stress data to determine values of k and n. The n is a measure of the deviation of the fluid from Newtonian behavior. If $n = 1$ then this expression simplifies to the Newtonian definition. When n is less than 1, it indicates the degree of "shear thinning"—the reduction in viscosity as shear rate increases. If the value of

n above 1, the opposite behavior-increasing viscosity with increasing shear rate is known as dilatancy.

3.2.5 Determination of Activation Energy of Flow

The effect of temperature on viscosity was evaluated using an Arrhenius-type Eq. (3.2) permitting determination of the activation energy for flow (ΔE_η) (Singh & Eipeson, 2000).

$$\eta = A \exp\left(" E_\eta / RT\right) \tag{3.2}$$

In Eq. (3.2), A is the temperature-independent pre-exponential term or the frequency factor, R is the gas constant (8.314 J/mol·K) and T is the reaction temperature in Kelvin. Taking the logarithm of Eq. (3.2), the plot of log(η) vs. $1/T$ should be linear with a slope of $(\Delta E_\eta/R)$. The liquid density was measured at 25 °C using a digital density/specific gravity meter (DA-110 M, Kyoto Electronics Manufacturing Co Ltd, Japan). The pH of the liquid was measured by a pH meter at 25 °C (Cyber Scan pH 1500, Eutech Instruments Pte Ltd., Singapore).

3.3 Production and Characterization of Alumina Support Particles

In this section, an attempt has been made to produce millimetric gamma-alumina particle using two different synthetic approaches, i.e., (a) integrated gelling process and (b) oil drop granulation process. Then, the properties of the particle produced by the two methods have been compared. Finally, the method that gave the better physical properties was chosen and used in subsequent experiments.

3.3.1 Materials

Sodium-alginate (Manugel GHB) with medium range molecular weight of 97,000, 37 % β-D-mannuronic acid residues (M) and 63 % α-L-guluronic acid residues (G) was supplied by ISP Technologies Inc, UK. Calcium chloride was purchased from Merck, Germany. Aluminum chloride ($AlCl_3$) was also supplied by Sigma-Aldrich, Germany. Ammonia solution was of analytical grade. Paraffin oil (0.870 g/mL density at 20 °C) was purchased from Ajax chemicals. Unmodified corn starch was obtained from Sigma-Aldrich Inc, USA. The water used in all experiments was deionized.

3.3.2   Measurement of Density, Surface Tension and Viscosity of Liquids

The liquid density was measured at 25 °C by using a digital density meter (DA-110 M, Kyoto Electronics Manufacturing Co Ltd, Japan). The liquid was manually pumped through a tube into a density meter and a stable reading of density was recorded. The drop weight method was used to measure surface tension of the liquid (Lee, Park, & Lee, 2009). According to the terminology proposed by Lee, Park, et al., 2009, this corresponds to a LCP (Lee-Chan-Pogaku) method. The drop weight results of a liquid sample were first obtained from dripping tips of different sizes (outer diameter from 0.30 to 1.65 mm). The data was then fitted into a quadratic regression equation in order to determine the linear coefficient of the equation. Finally, the surface tension of the liquid was determined by using a simple semi-empirical equation, $\gamma = 171.2\ C_2$ where C_2 is the second coefficient of a quadratic equation. The viscosity of the liquid was determined by using a rheometer according to standard procedure (Brookfield Engineering Laboratories, Inc., Model: DV III, USA) as described in Sect. 3.2.3.

3.3.3   Production of Particles Using Integrated Gelling Process

The millimetric particle was prepared according to the method described by Prouzet et al. (2006). 300 g of boehmite powder was dissolved into 1 L of deionized water, the suspension being dispersed under ultra-sound for 3 min at amplitude of 70 %. The boehmite suspension was then mixed with suspension of 20 g/L Na-alginate under magnetic stirring for 10 min, resulting suspension being left for 30 min. Subsequently, a mixture of alginate-boehmite suspension was dropped by pump through a needle (1.65 mm in diameter) into a gently agitated solution of 36.5 g/L $AlCl_3$ containing volume fraction of $CaCl_2$ ranged from 0.5 to 3.0 g/L, which were acidified at pH 1 by addition HCl. The air pressure of the pump regulated to ensure that the flow rate of the liquid was constant (0.4 mL/min) in all experiments. The other parameters used to prepare the particles are given in Table 3.2. The apparatus for particles formation is depicted in Fig. 3.4. Drops of the mixture suspension were

Table 3.2   The alginate-boehmite gel properties and particle preparation parameters

Parameters	Value
Viscosity, η (mPa.s)	321.12
Density, ρ (kg/m^3)	1127
Surface tension, (mN/m)	58.32
Tip size (mm in diameter)	1.65
Distance between the orifice of the needle and the surface of cross linking solution (cm)	10

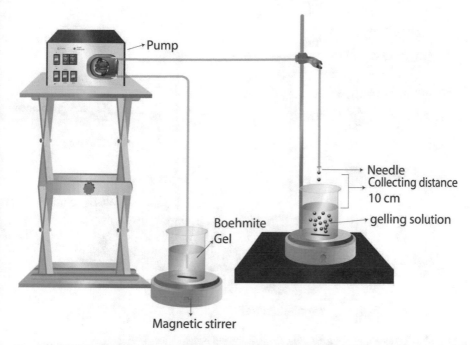

Fig. 3.4 Experimental setup for production of boehmite-alginate particles

progressively gelled and form particles that were left for aging in the gelifying solution consisting of AlCl₃ and CaCl₃ for 12 h. The beads separated by simple filtration were subjected to drying in the air after being washed with water. Finally, the particles were calcined at 300, 500 and 800 °C temperatures.

3.3.4   Production of Particles Using Oil Drop Granulation Process

Millimetric gamma-alumina (γ-Al_2O_3) support was prepared using oil drop granulation process previously described by Wang and Lin (1998). In brief, different amounts of boehmite powder (AlOOH) ranging from 13 g to 28 g were suspended (deionized water separately in 100 mL of deionized water), followed by addition of 2 g starch in each concentration under stirring. The suspensions were then dispersed ultrasonically for 3 min. at a amplitude of 70 %. The pH of the suspension was adjusted to 1.0 by the addition of about 4.5 mL of concentrated hydrochloric acid where the sol turned to a gel. The resulting mix gel was then transferred drop wise by a peristaltic pump into a liquid column consisting of paraffin oil in the upper layer and ammonia solution in the bottom layer. The schematic presentation

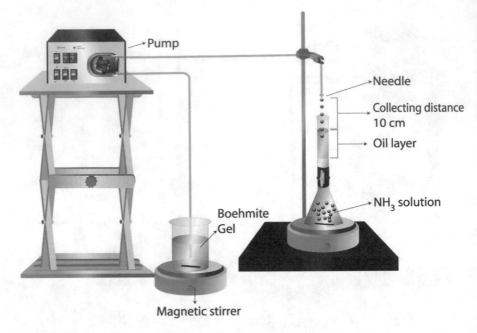

Fig. 3.5 Experimental setup for production of boehmite-starch particles

of the experimental system employed for production of the particles is shown in Fig. 3.5.

The air pressure of the pump regulated to ensure that the flow rate of the gel was constant (0.4 mL/min) in all experiments. The droplets were formed because of the surface tension effect during transiting through the oil layer and the gel droplets were aged in the ammonia solution for 1 h. During aging, the ammonia would neutralize the acid in the wet-gel particles and thus, the wet-gel droplets became rigid. Then, the particles were separated by simple filtration and washed with water. The particles were then dried in air at room temperature (25 °C) for 12 h and were later calcined at different temperatures 300–800 °C.

3.3.5 Determination of Particle Shape

In order to measure the shape of the particle quantitatively, sphericity factor (SF) was incorporated. Sphericity factor (SF) was used to express the divergence of a particle shape from spherical, as described by Chan (2011), where the value zero indicates a perfect sphere and higher values indicate a greater degree of shape distortion. The shape analysis of the beads were carried out by analyzing the digital images captured by a digital camera (Moticam-350, version 2.0 ML, China) installed on a stereozoom microscop (Stemi DV4, Carl Zeiss, Germany) using image analyzer (SigmaScan Pro 5.0, SPSS Inc). This program determines the D_{max} and D_{per} of the particle and computed using Microsoft Excel. SF was calculated according to Eq. 3.3.

$$SF = \left(D_{max} - D_{per}\right) / \left(D_{max} + D_{per}\right) \tag{3.3}$$

where, D_{max} is the maximum diameter passing through a beads centroid (mm) and D_{per} is the diameter perpendicular to D_{max} passing through the bead centroid (mm).

3.3.6   Determination of Particle Size

The size of particle was studied using Tate's law as described by Chan, Lee, Ravindra, and Poncelet (2009) from the following equation:

$$d_p = k\, k_{LF} \left(6d_T \gamma / \rho g\right)^{1/3} = K \left(6d_T \gamma / \rho g\right)^{1/3} \tag{3.4}$$

where, d_p is the overall diameter of bead (mm) and g is the gravitational force (m/s^2). γ and ρ are the surface tension (mN/m) and the density (kg/m^3) of the liquid, respectively.

The $k = k_g = k_a = k_c$ and k_{LF} defined as shrinkage factor and liquid lost factor respectively are calculated by Eq. (3.5) to (3.7) and Eq. (3.8), respectively;

$$k_g - d_g / d_f \tag{3.5}$$

$$k_a = d_a / d_f \tag{3.6}$$

$$k_c = d_c / d_f \tag{3.7}$$

$$k_{LF} = 0.98 - 0.04 d_T \tag{3.8}$$

where, k_g, k_a and k_c are the shrinkage factor of bead after gelling, after air-drying and after calcining respectively. $d_f d_g$, d_a and d_c are the diameter of the bead while falling, after gelling, after air-drying and after calcining, respectively. d_T is the dimensionless tip diameter (d_T) used in Eq. (3.8).

The progression of droplet injection at different stages was photographed by employing a digital camera (Canon, Japan) with a synchronized stroboscope light (Monarch, Nova-Strobe) having a frequency controller (100 to 14,000 flashes per minute). The resultant photographs of the bead were then transferred to image analyzer software (SigmaScan Pro 5.0, SPSS Inc). Diameter of the beads at different stages was measured directly from the image analyzer software, from which at least 100 measurements for each stage were analyzed to obtain an average value.

3.3.7   Determination of Surface Morphology of Particle

A scanning electron microscope (SEM, Model: JSM-5G1DLV, JEOL, Tokyo, Japan) was used to observe the surface morphologies of the particles. The SEM was an instrument that produces a largely magnified image by using electrons instead of light to form an image. It also had much higher resolution than traditional microscopes, because it uses electromagnets rather than lenses, allowing much more control in the degree of magnification. A beam of electrons was produced at the top of the microscope by an electron gun. The electron beam follows a vertical path through the microscope, which was held within a vacuum. The beam travels through electromagnetic fields and lenses, which focus the beam down toward the sample. Once the beam hits the sample, electrons and X-rays were ejected from the sample.

Detectors collect these X-rays, backscattered electrons, and secondary electrons and convert them into a signal that was sent to a screen. This produces the final image. The samples need to be made conductive by covering with a thin layer of platinum. This is done by using a device called a "sputter coater". As described by Peyrin, Mastrogiacomo, Cancedda, and Martinetti (2007), the sample is placed in a small chamber of a sputter coater in a vacuum which uses argon gas and a small electric field. Argon (Ar) gas is then introduced and an electric field is used to remove an electron from the argon atoms to make them ions with a positive charge. The Ar ions are then attracted to a negatively charged piece of platinum foil attached with sputter coater. Thus, the platinum atoms fall and settle onto the surface of the sample producing a thin platinum coating.

The cross-sectioned particles were mounted on metal stubs using double sided adhesive tape, dried under vacuum and spatter coated with platinum using a fine coater (JEOL, Japan, Model: JFC-1600) prior to SEM observation. The samples were mounted in the samples stub inside the scanning electron microscope and the images were viewed and captured under different magnifications.

3.3.8   Determination of Structure of Particle

X-ray diffraction analysis is the method most frequently used for the structural analysis of solids (Perego, 1998; Serwicka, 2000). A crystal lattice is a regular three-dimensional distribution (cubic, rhombic, etc.) of atoms in space. These are arranged so that they form a series of parallel planes separated from another by a distance, which varies according to the nature of the material. Figure 3.6 illustrates the X-rays scattered by atoms in an ordered lattice interfere constructively in directions given by Bragg's law. The XRD pattern of a powdered sample is measured with a stationary X-ray source (usually Cu Kα) and a movable detector, which scans the intensity of the diffracted radiation as a function of the angle 2θ between the incoming and the diffracted beams.

X-rays

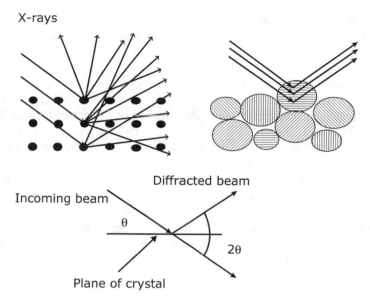

Fig. 3.6   The X-rays scattered by atoms in an ordered lattice interfere constructively in directions given by Bragg's law. Source: Greenberg (1989)

In this study, the structure of particles was determined using a Shimadzu diffractometer, model XRD-6000. The diffractometer employed Cu-Ka radiation to generate diffraction patterns from powder crystalline samples. All samples were mounted on sample holders, and the measurements were performed at 2θ values between 10 and 70°, with a step size of 0.05° at a speed of 0.05 s^{-1}. The data were analyzed with the DiffracPlus software and the phases were identified using the powder diffraction file (PDF) database (JCPDS, International Centre for Diffraction Data).

3.3.9   Determination of Basicity of Particle

Temperature-programmed desorption (TPD) analysis is a well established technique for characterizing the basic sites on the heterogeneous catalysts. The analysis involves heating a sample while contained in a vacuum and simultaneously detecting the residual gas in the vacuum by means of mass analyzer (Bertgeret & Gallezot, 1997). As the temperature rises, certain adsorbed species will have enough energy to desorb from the surfaces of the catalyst. The temperature of the desorption peak maximum is indicative of the strength with which the adsorbate is bound to the surface (Knozinger, 1997). The higher the temperature of the desorption peak the stronger the bond between the adsorbate and the surface. Temperature-programmed desorption uses probe molecules to examine the interactions between the surface with gases molecules (Knozinger, 1997). The probe molecules are chosen with respect to the nature of the adsorbed species believed to be important for the catalytic reaction under study or chosen to provide information about the specific type of surface sites, as described by Knozinger (1997).

Temperature programmed desorption with CO_2 as a probe molecule was applied in the study to measure the basicity of the catalyst as described by Yap, Lee, Hussein, and Yunus (2011). The TPD-CO_2 experiments were performed using a Thermo Finnigan TPD/R/O 1100 series catalytic surface analyzer equipped with a thermal conductivity detector. Catalysts (100 mg) were pretreated under a helium stream at 800 °C for 30 min (10 °C min^{-1}, 30 mL min^{-1}).

Then, the temperature was decreased to 30 °C, and a flow of pure CO_2 (30 mL min^{-1}) was introduced into the reactor for 1 h. The sample was flushed with helium at 30 °C for 30 min prior to the CO_2 desorption analysis. The analysis of CO_2 desorption was then carried out up to 800 °C under a helium flow (10 °C min^{-1}, 30 mL min^{-1}), and the amount of desorbed CO_2 was detected and determined using a thermal conductivity detector.

3.3.10 Determination of Surface Area, Pore Size Distribution and Pore Volume of Particle

The adsorption–desorption isotherms were determined by introducing measured volumes of the adsorbate to the sample bulb and measuring the equilibrium pressure. The adsorption data was plotted as the specific volume adsorbed, V_{ads} (cm^3/g), as a function of the relative vapor pressure $\dfrac{P}{P_o}$, where P is the equilibrium pressure and P_o is the saturated vapor pressure of nitrogen. One of the most frequently used models to determine the surface area of porous solids was the BET-model, established by Brunauer, Emmett and Teller, which was written in its linear form as (Gregg & Sing, 1982):

$$\frac{P/P_0}{V_{ads}\left(1-P/P_0\right)} = \frac{1}{V_m C} + \frac{C-1}{V_m C}\cdot\frac{P}{P_0} \qquad (3.9)$$

Where; V_{ads} is the amount of adsorbed gas, V_m is the amount of adsorbed gas at one monolayer, C is the equilibrium constant of adsorption in the first adsorption layer at the measuring temperature.

By plotting the experimental data in the form of $\dfrac{P/P_0}{V_{ads}\left(1-P/P_0\right)}$ versus $\dfrac{P}{P_0}$, a straight line was obtained; by linear regression the straight line intercept, $\dfrac{1}{V_m C}$, and slope, $\dfrac{C-1}{V_m C}$, was evaluated. From the equations for slope and intercept, the values of V_m and C were obtained separately. The BET-surface area of the sample was obtained using the following equation:

$$S_{BET} = V_m \cdot \sigma \cdot N_A \qquad (3.10)$$

Where; σ is the molecular cross sectional area for nitrogen (=0.162 nm2) and N_A is Avogadro's number, the BET-surface area is obtained.

BJH (Barrett, Joyner, and Halenda) method was the most widely used method to calculate the pore size distribution (Storck, Bretinger, & Maier, 1998; Leofanti, Padovan, Tozzola, & Venturelli, 1998). This method involves an imaginary emptying of condensed adsorptive in the pores in a stepwise manner as relative pressure decreasing. Pore size distributions are obtained from BJH fits of the N_2 desorption isotherms. This method is based on Kelvin's equation, stating that condensation occurs in pores with radius r_m at a relative pressure $\dfrac{P}{P_o}$, which is represented by;

$$\ln \frac{P}{P_0} = -\frac{2\gamma V_m \cos\theta}{RTr_m} \qquad (3.11)$$

where; γ is the liquid surface tension of the liquid–vapor interface, V_m is the molar volume of condensable vapor, θ is the contact angle between the solid and the condensed phase. R is the universal gas constant and T is the absolute temperature. A widely applied technique, Gurvich method, introduced by Rouquerol, Rouquerol, and Sing (1999) was used to determine pore volume of the catalyst. The adsorbate has the same volume as an equivalent quantity of bulk liquid is known as the Gurvich rule. The pore volume V_{pore} (cm³/g) is given by;

$$V_{pore} = V_{ads} \Big/ \rho \qquad (3.12)$$

Where; V_{ads} is the adsorbed amount of gas, ρ is the liquid density.

In the work, the surface area, pore size distribution and pore volume was measured using an an automatic adsorption instrument (Sorptomatic 1990, Thermo Finnigan Italia S.p.A). The analysis was conducted using Thermo Finnigan Sorptomatic 1900 series nitrogen adsorption–desorption analysis software. All of the samples were degassed at 150 °C under vacuum conditions until no pressure gradient could be detected as reported by Yap et al. (2011).

3.3.11   Determination of Crush Strength of Particle

The crushing strength of the individual beads was determined by texture analyzer (TA.XT. Plus, Texture Technologies, USA). The particle is placed at the lower plate and the upper plate is lowered gradually until it breaks the particle and the maximum force applied to break the single particle was taken as crush strength. Four individual particles which have the same diameter were tested for each sample.

3.4   Production and Characterization of Alumina Beads Supported Catalyst

3.4.1   Model Catalysts

Potassium iodide, KI (Sigma-Aldrich, GmbH, Germany), sodium nitrate, $NaNO_3$ (Sigma-aldrich), and potassium fluoride, KF (VWR international) were used as model catalysts.

3.4.2   Preparation of Alumina Beads Supported Catalyst

The alumina particles were produced by the integrative gelling process or oil-drop granulation process as described in Sects. 3.3.3 and 3.3.4 respectively. The catalysts were loaded on the alumina beads by impregnating the aqueous solutions of catalysts for 1 h. The particle-supported catalysts were then calcined in a furnace at from 500 °C to make the catalyst active for transesterification reaction. The amount of KI KF and $NaNO_3$ impregnation was maintained 0.06, 0.15, 0.24, 0.30, and 0.33 g (catalyst/$g_{\gamma-Al2O3}$). The basic principle of impregnation methods is the amount of support is equal to the amount of solution absorbed by the support. The impregnation of catalyst is given in Table 3.3.

Table 3.3   Impregnation of millimetric catalyst

Step-1	Step-2		Step-3	Step-4		Step-5
Catalyst (g) for 1 g support	Catalyst (g) for 3 g of support	Total weight of catalyst and support (g)	Volume of aqueous solution (ml) with catalyst and 3 g of support was immersed	Calcination	Total weight (g)	The calcinated catalyst, g (catalyst+support) used for biodiesel production (4 wt%, g cat./g oil)
0.06	3×0.06=0.18	3.18	4.5	Calcination of impregnated catalyst	>2.5	0.6
0.15	3×0.15=0.45	3.45	4.5			0.6
0.24	3×0.24=0.72	3.72	4.5			0.6
0.30	3×0.30=0.90	3.90	4.5			0.6
0.33	3×0.33=0.99	3.99	4.5			0.6

Table 3.4 Characterizations of particle-supported catalyst

Properties of particle-supported catalyst	Characterization
Shape	Sphericity factor
Size	Tates Law
Surface morphology	Scanning electron microscope (SEM)
Crystal structure	X-ray diffraction (XRD)
Basicity	Temperature-programmed desorption of CO_2 (CO_2-TPD)
Surface area and pore structure	N_2 adsorption–desorption isotherms

3.4.3 Characterizations of Particle-Supported Catalyst

Characterizations of gamma alumina particle supported catalyst are summarized in Table 3.4 and the methods are described in Sects. 3.3.5 to 3.3.10.

3.5 Production and Analysis of Biodiesel

3.5.1 Materials

Commercial edible grade palm oil (Buruh, Lam soon edible oil, SDN, BDH, Selangor, Malaysia) was purchased from the supermarket. Methanol (Labscan Asia Co., Ltd., reagent grade 99.9%) was used for the transesterification reactions. Normal hexane (Fisher scientific, analytical reagent grade), methyl heptadecanoate (Fluka analytical) and other standards (methyl myristate, methyl palmitate, methyl stearate, methyl oleate, and methyl linoleate) for GC column calibration were purchased from Sigma Chemical Co.

3.5.2 Transesterification Reaction Conditions

The transesterification was carried out in a 50 ml baffled conical flask (Corning, USA) placed in an incubator orbital shaker equipped with a temperature controller. The catalysts were calcined before use for the reaction. The baffled conical flask was charged with of palm oil (15 g), varied amount of methanol and catalyst with different amount of KI, KF, and $NaNO_3$ loading. Each reaction was performed at 60 °C with shaking at a speed of 250 rpm for the required time.

After completion of the reaction, the catalyst was recovered from the reaction mixture by simple filtration. The filtrate was then centrifuged at a relative centrifugal force of $2500 \times g$ for 10 min, and then excessive amount of methanol was

Table 3.5 Process variables and optimization of biodiesel production

Process variables[a]	Range of tested variables
Effect of catalyst loading on FAME yield	0.06 to 0.33 g(catalyst)/g$_{\gamma\text{-Al2O3}}$
Effect of reaction time on FAME yield	1–6 h
Effect of methanol/oil molar ratio on FAME yield	8–18

[a]Each reaction was performed at 60 °C with shaking at a speed of 250 rpm for the required time in a batch process

evaporated before analysis of biodiesel yield. The process variables and the range of tested variables are given in Table 3.5.

3.5.3   Analysis of Biodiesel

Gas chromatography is commonly adopted to determine the fatty acid methyl ester (FAME) yield (%) according to the EN 14103 standard method (Alonso, Mariscal, Tost, Poves, & Granados, 2007). EN 14103 calls for calibration of all FAME components by relative response to a single compound, methyl heptadecanoate. This requires the measurement of accurate weights for each sample and the addition of an internal standard. The response factor (RF) is essentially a measure of the relative response of the instrument detector to an analyte compared to an internal or external standard. The RF can be calculated according to the EN 14103 using the following equation:

$$\text{RF}_{\text{RS}} = \frac{A_{\text{is}} \times C_{\text{RS}}}{A_{\text{RS}} \times C_{\text{is}}} \tag{3.13}$$

Where,	
A_{is}	Area of internal standard
C_{RS}	Concentration of reference standards (methyl myristate, methyl palmitate, methyl stearate, methyl oleate, and methyl linoleate) in solution
A_{RS}	Area of reference standard
C_{is}	Concentration of internal standard (methyl heptadodecanoate) in reference standard solution

Using the response factors, the amount of each fatty acid (as corresponding methyl ester), F_{TG}, in test sample was computed as follows:

$$F_{\text{TG}} = C_{\text{iss}} \times A_{\text{FCS}} \times \text{RF}_{\text{RS}} / A_{\text{iss}} \tag{3.14}$$

Where,	
C_{iss}	Concentration of internal standard in the sample
A_{iss}	Area of internal standard (methyl heptadodecanoate) in the sample
A_{FCS}	Area of individual FAMEs compound in the sample
RF_{RS}	Response factor of the respective reference standard.

Calculate the amount of FAME Yield (%) in the test sample as follows:

$$\text{FAME Yield}(\%) = \frac{\text{Amount of FAME produced experimentally}(\text{mole})}{\text{Amount of FAME produced theoretically}(\text{mole})} \times 100$$

(3.15)

Calculate the amount of FAME produced theoretically (expressed as mole) in test sample as follows:

$$\text{Amount of FAME produced theoretically}(\text{mole}) = \frac{\text{Amount of oil used for reaction}}{\text{Molecular weight of palm oil}} \times 3$$

(3.16)

In this experiment, the biodiesel product was analyzed by gas chromatography (Shimadzu GC-14B) using flame ionization detector (FID), equipped with a capillary column (30 m×0.5 mm×0.25 μm). The injector and detector temperatures were 240 and 280 °C, respectively. The initial oven temperature was 80 °C, with an equilibrium time of 1 minute. After the isothermal period, the oven was heated at a rate of 10 °C/min to 270 °C, and the temperature was maintained for 7 min. Hexane was used as a solvent and methyl heptadecanoate was used as an internal standard. In all, 0.2 g of biodiesel sample was weighed and added to a solution of methyl heptadecanoate at a concentration of 0.1 g per 100 mL of hexane. A total of 0.5 μL of the resulting solution was injected into the gas chromatograph, and helium was used as the carrier gas. The gas chromatogram of the biodiesel was shown in Fig. 3.7.

3.6   Reusability of Catalyst

Catalyst reusability test were performed by repeating the transesterification over an extended period of time with used catalysts. When the transesterification reaction was completed, the catalyst was recovered by filtration and dried at 100 °C for 1 hour and subsequently reused. The same amount of palm oil and methanol were added to the recycled catalyst each time to react under the same conditions. This procedure was repeated several reaction cycles to examine the extent of stability of the catalyst.

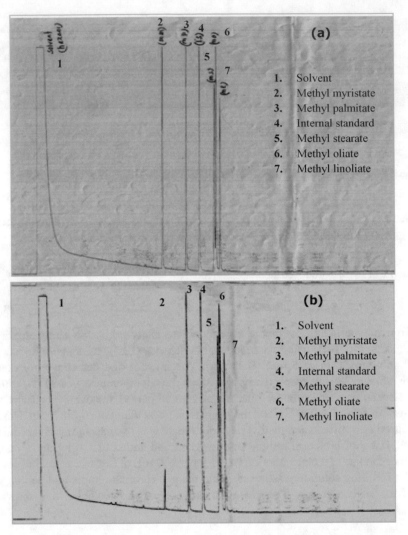

Fig. 3.7 Gas chromatogram of (**a**) standard methyl ester and (**b**) experimental biodiesel product

3.6.1   Determination of Chemical Composition of Reused Catalyst Particle

Elemental analysis on the reused catalyst was carried out by X-ray fluorescence in a Shimadzu EDX 720 spectrometer with a rhodium X-ray source tube. The sample was placed on powders placed inside the sample holder using a polypropylene film. The analysis were performed under vacuum conditions (<45 Pa) using 2-channels where the X-ray source was set at 50 and 15 kV for Titanium (Ti)–Uranium (U) and Sodium (Na)–Scandium (Sc) ranges, respectively and the metal oxide content of the catalysts was measured automatically by the instrument software.

Chapter 4
Development of Millimetric Particle for Biodiesel Production

4.1 Introduction

This chapter includes the effect of pH on the gelling behavior, viscosity, rheology, and density of the boehmite suspension prior to the development of millimetric gelled boehmite particles. The shape and size of gelled particles produced from both processes were evaluated using sphericity factor and Tates Law, respectively. In addition, the characteristics of the synthesized particle, such as the structural, morphological, textural properties, were carried out using X-ray diffraction (XRD), scanning electron microscopy (SEM), N_2 adsorption–desorption isotherms. The shrinkage (%) of the particle at different preparation steps and the mechanical strength of the particle were also evaluated. The comparison of the particle properties produced from two methods was made to select the suitable method in order for it to be used in subsequent works.

4.2 Gelling Behavior of the Boehmite Suspension

The boehmite particles of the order of micrometers (Fig. 4.1) were suspended in water at different concentrations (5–30 %w/v). The gelling as well as the rheological behavior of the boehmite suspension was determined with respect to different pHs of the boehmite suspensions.

© Springer International Publishing Switzerland 2017
A. Islam, P. Ravindra, *Biodiesel Production with Green Technologies*,
DOI 10.1007/978-3-319-45273-9_4

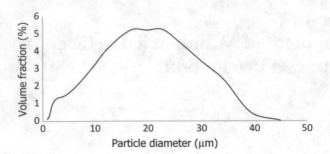

Fig. 4.1 The particle size distribution of boehmite powder

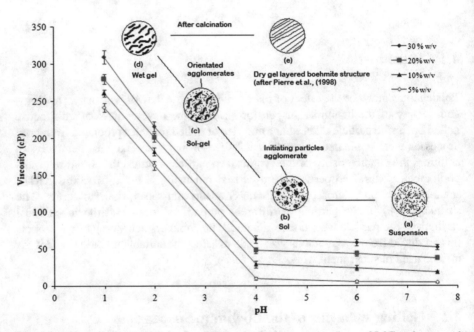

Fig. 4.2 Effect of pH on viscosity for different boehmite concentration at 25 °C and proposed gelling model depicting the morphological changes occurring in a boehmite suspensions at controlled pH of 1–7.6

4.2.1 Effect of pH on Viscosity

Figure 4.2 shows the effect of pH on the viscosities of boehmite suspensions. The viscosities of the suspensions at various boehmite concentrations (i.e., 5–30 % w/v) were found to be relatively constant over a pH range of 4–7.6. When the pH was lower than 4, a remarkable increase in viscosity was observed for all boehmite concentrations due to the formation of a boehmite gel with a greater density. Increases in the viscosities with decreasing suspension pH were more remarkable for higher concentrations.

The lowest viscosity values were observed from pH 7.6–4, indicating that in this region the boehmite particles are well dispersed at all concentrations. With the decrease in pH from 4.0–1.0, the viscosity then increased by several orders of magnitude, which could be indicative of a tendency of boehmite particles to change from the dispersed state to the aggregated state due to the weak attractive forces (Yoldas, 1975).

The gelling behavior can be explained as, the pH of the suspension is decreased, individual boehmite particles are charged by hydrogen ions which are released by the dissociation of hydrochloric acid. At the same time, the positively charged boehmite particle can absorb counter ions (chloride ions of the acid) neutralizes the surface charge, effectively decreasing the interparticle repulsion Thus, the particles tends to come closer due to weak forces (e.g., polar and van der Waals forces) as the attractive force became predominant over the repulsive forces at low pH and aggregates are formed and water is probably trapped by them. This results in increase in viscosity. Moreover, the increase of the attractive forces among the particles implies higher viscosity until gelation is complete. In addition, the rapid development of particle-particle interactions may take place as pH is decreased (Cristiani, Valentini, Merazzi, Neglia, & Forzatti, 2005). Figure 4.1 shows a sharp increase in the viscosity of the suspensions at pH 1 as the gelation point is approached.

4.2.2   Proposed Model

Based on the results of the effect of pH on viscosity, a schematic model indicative of the rheological changes of boehmite suspensions over the range of pH studied is presented in Fig. 4.2. At pH 7.6, the boehmite particles are suspended and unevenly distributed throughout the suspension (Fig. 4.2a). Decreasing the pH from 7.6 to 4.0 results in the formation of several aggregates as depicted in Fig. 4.2b potentially due to the weak forces of attraction described in the former section. The boehmite particles in these aggregates may not be in actual contact with one another due to the strong repulsive forces between the positive surface charges imparted to the particles by acid treatment (Song & Chung, 1989). As the pH of the suspension decreased, the primary particles in the suspension tend to attract one another and are then likely to gather into aggregates as shown in Fig. 4.2c. pH values of less than 4 in the boehmite suspension would favor the formation of a partially oriented gel structure (Fauchadour, Kolenda, Rouleau, Barre, & Normand, 2000). Moreover, the orientation of the structured gel depends on the charge distribution on the primary particles determined by the electrolyte concentration in solution (Song & Chung, 1989).

The network structure of Fig. 4.2d could be formed under more strongly acidic conditions that result in a strongly preferred growth direction in agglomerates which becomes more pronounced with increasing acid content. This result is in a good agreement with the results of previous work (Rueb & Zukoski, 1992). The final stage in the wet-gel synthesis is referred to as the "post-gelation" step. Changes that occur during the drying and calcination of the wet gel include the desorption of water, the evaporation of the solvent, and structural changes. The formation of the layered structure of wet-gel boehmite after calcination may be related to the parallel stacking of particles due to the phase transition from a random to an ordered state (Popa, Rossignol, & Kappenstein, 2002) as shown in Fig. 4.2e. This study provided evidence for the formation of a dry-gel layered structure of boehmite as reported by Yoldas (1975).

Figure 4.3 shows that all the Al atoms are in octahedral coordination due to their higher ionic character and denser packing about an Al atom. As boehmite particles have a layered structure somewhat similar to that of clays, interparticle bonding is achieved by weak hydrogen bonds (Li, Smith, Inomata, & Arai, 2002). Therefore, to achieve a gel in a boehmite suspension, it may be necessary to strengthen the initial hydrogen bonds between lamellar particles. As a consequence, a continuous decrease in pH may bring the particles closer, resulting in a stronger interaction and finally sufficient development of a layered structure.

4.2.3   Sedimentation Study

Figure 4.4 presents photographs of the settling behavior of the boehmite suspensions at different pHs. A clear settling zone could be observed at pH values above 4, whereas a more homogenous gelling phase was observed at pH 4.0 and below. Philipse (1993) have discussed the stability of suspension for the case of boehmite silica mixture and suggested that the stable suspension was formed due to the attraction force between particles under acidic condition. Thus, the absence of settling at pH 1 and 2 (Fig. 4.4) could be attributed to an increase in density due to flocculation effect between particles. According to Stoke's law, the increase in viscosity could also be involved.

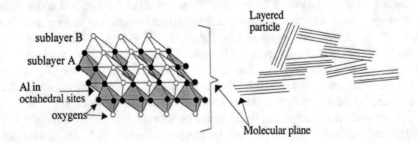

Fig. 4.3 Boehmite gel structure Source: Pierre et al., (1998)

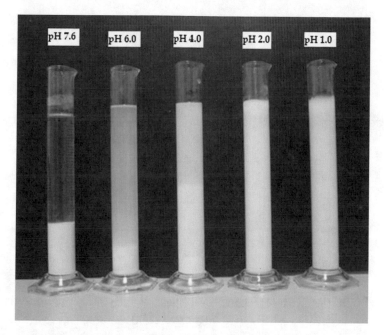

Fig. 4.4 Settling behavior of 30 %(w/v) boehmite suspensions at different pH values

4.2.4 Effect of Shear Rate on Viscosity of the Suspension

Figure 4.5 presents typical flow curves for the boehmite suspension at different pH values. The viscosity decreases when the shear rate is increased from 10 to 100 s⁻¹, thus confirming the pseudoplastic behavior of the suspensions. Strong shear dependence is observed for all suspension pH values investigated here. This can be explained by the fact that the associations between particles are formed just after lowering the pH and that the breaking up of those aggregates at high shear rates results in shear thinning behavior. Figure 4.5 shows a region at high shear rates (>150 s⁻¹) where the viscosity is nearly independent of the shear rate. The viscosity in this region attains the "low Newtonian limit" of the suspension.

However, at the low shear rate, the structure oriented by shear returns to the initial structure, and the particles easily produce more "cardhouse" structures that can enhance the viscosity of the suspension (Song & Chung, 1989).

The power law model was found to give a close fit to the experimental data (Fig. 4.6) with correlation coefficient (R) values greater than 0.99 (Table 4.1). The values of n less than 1 indicative of the deviation of flow from the Newtonian behavior (Chhinnan, Mcwatters, & Rao, 1985) were dependent on pH of the suspensions used as shown in Table 4.1.

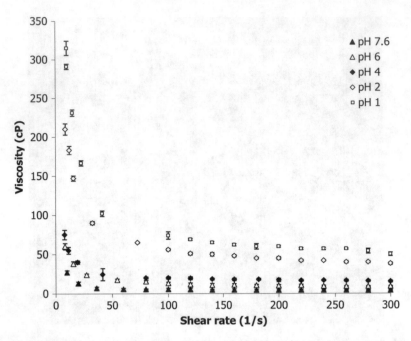

Fig. 4.5 Relationship between shear rate and viscosity of 30%(w/v) boehmite suspensions for different pHs at 25 °C

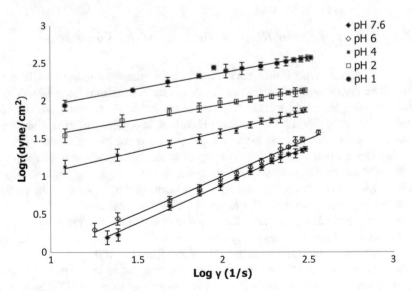

Fig. 4.6 Shows the Power Law plots using the shear rate vs. data for pH from 1.0 to 7.6 at 30% (w/v) of boehmite suspension at 25 °C

Table 4.1 The flow characteristics parameter estimated from the shear rate and shear stress data of 30 % (w/v) boehmite suspension prepared from various pH at 25 °C

pH	K (Pa.s)	n	R^2
1.0	2.978	0.398	0.992
2.0	1.432	0.443	0.987
4.0	0.414	0.494	0.992
6.0	0.010	0.988	0.993
7.6	0.007	0.994	0.996

The low n values obtained from the low pH, suspensions represent a significant departure from Newtonian flow behavior and these suspensions have high viscosities at low shear rates that decrease dramatically as the shear is increased. These relatively low n values indicate either a continuous breakdown or a realignment of the molecules of the suspensions along the direction of flow (Mikolajczyk, Czapnik, & Bogun, 2004). Moreover, a high shear rate can facilitate the molecules straightening, causing the liquid to assume a more non-Newtonian character. These phenomena are characteristic of fluids rarefied by shear in which the suspension with high viscosity become less viscous or less dense due to the straightening molecules along the flow direction during the action of shear stress. The consistency index k increases with decreasing suspension pH. The high value of k is mainly attributed to the growing intermolecular friction forces and interactions. Moreover, the net charge of boehmite molecules at low pH may cause an expansion of the boehmite molecules, accompanied with higher hydration and swelling that may increase the value of k. Furthermore, the correlation between the rheological parameters k and n demonstrates that these parameters showed no consistent trend with pH and concentration.

4.2.5   The Effect of Temperature on Viscosity and Flow Activation Energy

Figure 4.7 plots the natural logarithm of the apparent viscosity against the reciprocal of the absolute temperature (K). The activation energies of the flow processes were calculated according to Eq. (3.2), and the results are summarized in Table 4.2. Temperature was found to have a considerable effect on the viscosities of boehmite suspensions, with a general tendency of decreasing viscosity with increasing temperature. The activation energy (ΔE) of the boehmite suspension (30 % w/v) decreased with increasing solution pH from 1 to 7.6 (Table 4.2).

Temperature relationship with viscosity based on Arrhenius type equation gave activation energy (ΔE) in the range of 13.41 kJ/mol at pH 1–5.47 kJ/mol at pH 7.6 (Table 4.2). It can be inferred from Table 4.2 that the suspensions with the high pH showed high sensitivity to temperature with low activation energy. These results are also in agreement with previously published studies (Feng, Gu, & Jin, 2007) which presented identical outcomes. The activation energy can be correlated with the

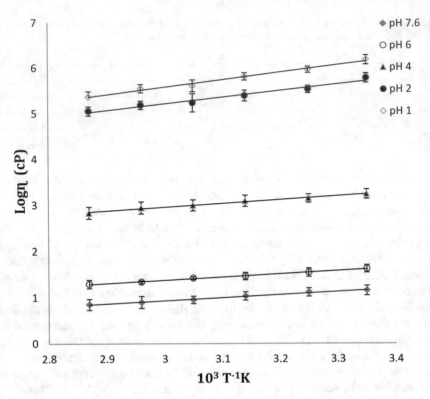

Fig. 4.7 Viscosity of boehmite suspension of 30%(w/v) at different pHs as a function of temperature

Table 4.2 Influence of pH on gelation at various temperatures of 30%(w/v) boehmite suspensions

pH	Activation energy (ΔE) (J/mol)	R^2
1	13.41	0.989
2	11.88	0.981
4	6.71	0.987
6	5.69	0.994
7.6	5.47	0.989

Table 4.3 The porosity and mechanical properties of particles

Parameters	Air-dried particles	Calcined particles
BET surface area (m²/g)	119	244
Average pore diameter (nm)	8.40	10.60
Pore volume (cm³/g)	0.44	0.83
Mechanical strength (N/particle)±S.D.	5±1.0	3±0.4

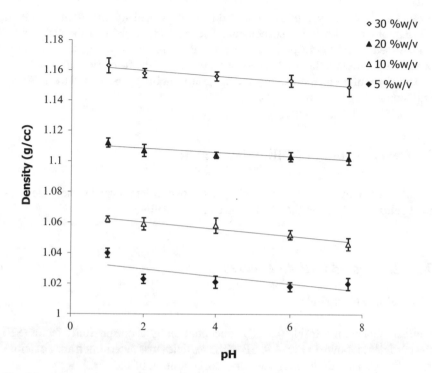

Fig. 4.8 Effect of pH on the density of different concentrations boehmite suspensions at 25 °C

energy required for breaking the chemical bonds which facilitates faster flow, as reported by Poulain et al. (1996). The higher values of ΔE of our sample at low pH can be explained in terms of an increase in the interparticle attraction forces between particles that tend to draw the particles together. The lower values of ΔE can be interpreted by the weak attractive forces between boehmite particles at higher pH.

4.2.6   Effect of pH on Density

The pH of the boehmite suspension also affected the density of the boehmite suspensions as shown in Fig. 4.8. The difference in density was more robust in suspensions with higher boehmite concentrations. The boehmite gel density has been reported to vary as a function of pH due to the precipitation to the agglomerate state (Maskara & Smith, 2005). Moreover, the interparticle force at low pH might be responsible for increasing the density of the boehmite gel. The density of the boehmite suspensions will vary depending on pH, with basic suspensions having a lower density than acidic suspensions.

   From the above study, it can be inferred that the pH had significant influence on gel formation of the boehmite suspension. The viscosity, density of the suspension was increased sharply in response to a decrease in pH to 1 as the gelation point is approached. Therefore, the pH 1 could be favorable for the formation of gelled particle using two different approaches; the integrative gelling process and the oil-drop granulation gelling process.

4.3 Development of Millimetric Particle

The millimetric particle was developed by using two different approaches; the integrative gelling process and the oil-drop granulation gelling process.

4.3.1 Integrated Gelling Process

4.3.1.1 Shape of Particles

The effect of calcium chloride ($CaCl_2$) concentration (g/L) on sphericity factor (SF) of the particles is shown in Fig. 4.9. SF of the particles was determined after calcination at 800 °C. The results indicated that concentration of the $CaCl_2$ up to 2.5 g/L caused decrease in the SF. Increasing the $CaCl_2$ concentration higher than 2.5 g/L, however, caused no significant change (SF<0.03) in the sphericity factor of particles. It can thus be inferred that the $CaCl_2$ concentration of 2.5 g/L could be sufficient for spherical alginate-boehmite particle formation.

   The photographs of typical alginate-boehmite particles obtained with different $CaCl_2$ concentrations ranging from 0.0 to 3.0 g/L, are shown in Fig. 4.10a-g. It has

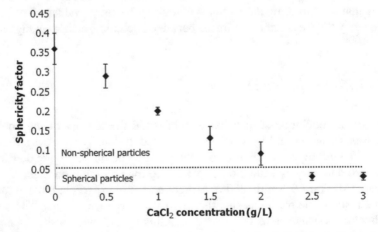

Fig. 4.9 Effect of $CaCl_2$ concentration on sphericity factor of the alginate-boehmite particles

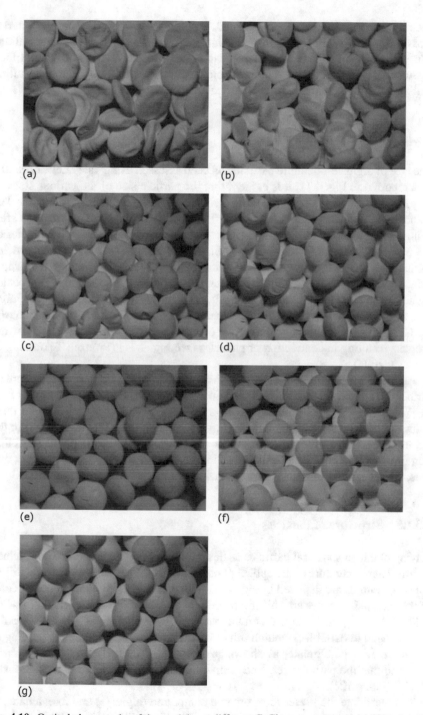

Fig. 4.10 Optical photographs of the particles at different $CaCl_2$ concentrations. (a) Without adding $CaCl_2$, (b) 0.05 g/L $CaCl_2$, (c)1.0 g/L $CaCl_2$, (d) 1.5 g/L $CaCl_2$, (e) 2.0 g/L $CaCl_2$, (f) 2.5 g/L $CaCl_2$, (g) 3.0 g/L $CaCl_2$

been reported by Morch, Donati, Strand, and Skjak-Baek (2006) that the interior cross-linking between the alginate chains was accompanied by divalent cations (Ca^{+2}). Thus, the addition (2.5 g/L) of salt ($CaCl_2$) to the gelling bath would then presumably lead to sufficient internal ionic cross-links within a 3D-lattice resulting in the instantaneous formation of spherical-shaped particles.

4.3.1.2   Size of Particles

The size of alginate-boehmite particle has been assessed using Tates Law. From the data reported in Fig. 4.11, it is evident that the particle size of the various steps of preparations was obtained in the range between 3.40 and 1.98 mm in diameter. The reduction of diameter after gelling was 10 % whereas it was increased to 42 % after being calcined at 800 °C (see Fig. 4.11). After gelation, the decreased diameter of the particles could be explained by gelling mechanism associated with particles formation process. The alginate-boehmite gelling process consists of two steps-gelation of a biopolymer, sodium alginate, followed by gelation of mineral species such as boehmite (Prouzet et al., 2006). It has been stated that when a drop of alginate solution comes in contact with calcium ions, gelation occurs instantaneously. As Ca^{2+} ions penetrate into the interior of droplets, water is squeezed out of the droplets resulting in contraction of particles (see Fig. 4.11) (Prouzet, Tokumoto, & Krivaya, 2004).

After the process is completed, condensation of inorganic part of the droplets further occurred once the droplets were in contact with the acidic pH (pH less than 1) (Yoldas, 1975). This process was also associated with water release and thus, reduced the volume of the particle. However, when the particles were dried in the air, it was found that the particles shrank significantly (36 %). Finally, the calcination of particles could lead to the cohesion of mineral (boehmite) framework and thus, the reduction of particles size reaches a maximum (42 %) (see Fig. 4.11).

4.3.1.3   Structure of Particles

Having obtained spherical particles as described above, the particles were calcined at different temperatures. The optical photographs of the particles calcined at different temperatures are depicted in Fig. 4.12a-c. It can be seen that the modes of color of the calcined particles are shifting from light gray at 300 °C (Fig. 4.12a) to white color at 800 °C (Fig. 4.12c). When the particles were calcined at 500 °C, the color was changed to dark black, which might be attributed to the incomplete burning of organic materials, alginate, as shown in Fig. 4.12b. Furthermore, it was clearly observed that the particles retained their spherical shape even after calcinations at 800 °C (Fig. 4.12c).

To investigate the presence of organic compound (alginate) and to evaluate the structure of the particles after calcinations, the X-ray diffraction (XRD) analysis was performed (Fig. 4.13). The X-ray diffraction patterns of the particles (Fig. 4.13a)

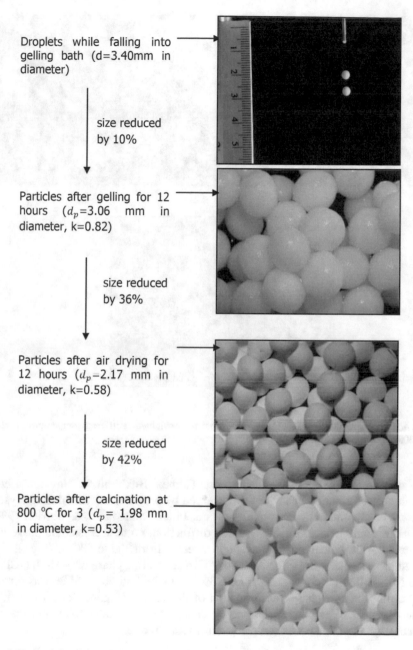

Droplets while falling into gelling bath (d=3.40mm in diameter)

size reduced by 10%

Particles after gelling for 12 hours (d_p=3.06 mm in diameter, k=0.82)

size reduced by 36%

Particles after air drying for 12 hours (d_p=2.17 mm in diameter, k=0.58)

size reduced by 42%

Particles after calcination at 800 °C for 3 (d_p= 1.98 mm in diameter, k=0.53)

Fig. 4.11 Particles size at different preparation steps

following calcination at 300 °C exhibited the characteristic peaks of Al_2O_3 (JCPDS File No. 00-031-0026) at 2θ: 19.58, 32.77, 37.60, 39.49, 44.53) and carbon (JCPDS File No. 00-026-1077) at 66.76 of 2θ. The main reason for the transformation to Al_2O_3 from AlO(OH) was associated with the dehydration of the sample when it

Fig. 4.12 Optical photographs of the spherical particles calcined at different temperatures for 3 h. (**a**) 300 °C, (**b**) 500 °C, and (**c**) 800 °C

was calcined at 300 °C (Vazquez, Lopez, Gomez, Bokhimit, & Novarot, 1997). Similar transformation behavior was observed by previous workers during crystallization of aluminum hydrous oxides (Mani, Pillai, Damodaran, & Warrier, 1994). As far as could be determined by X-ray diffraction, no new crystalline phase was observed as the calcination temperature increases from 300 to 500 °C (Fig. 4.13b). When the particles were calcined at 800 °C, a crystalline phase of γ-Al_2O_3 (JCPDS File No. 00-029-0063) at 31.93, 37.60, 39.49, 45.79, and 66.73 of 2θ was formed suggesting the new phase transformation of the sample (Fig. 4.13c). However, no reflection peak of carbon species was observed. It is reasonable to assume that the combustion of alginate was complete at this temperature.

4.3.1.4 Surface Area and Porosity of Particles

The nitrogen adsorption–desorption isotherms, recorded for both of the air-dried and the calcined particles are displayed in Fig. 4.14. These isotherms possess features reminiscent of IV type according to the IUPAC classification, exhibiting

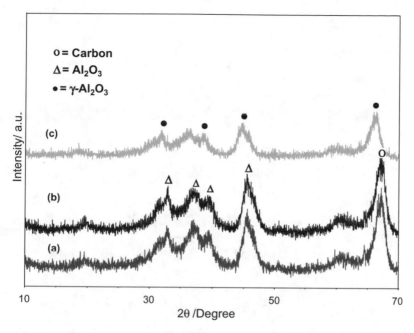

Fig. 4.13 XRD patterns of the sample after calcination for 3 h at various temperatures: (**a**) 300 °C, (**b**) 500 °C, (**c**) 800 °C

characteristic hysteresis loops. The major uptake observed at high relative pressures (P/P_0) range 0.62–0.80 for particles dried with air (Fig. 4.14a) and at 0.41–0.85 for the calcined particles (Fig. 4.14b). According to the theoretical considerations, the linear portion of the curve represents multilayer adsorption of nitrogen on the internal surface of the sample, and the concave upward portion of the curve represents filling up of mesopores (2–50 nm) and macropores (>50 nm), which would call forth the capillary condensation process.

Table 5.1 summarizes the total BET surface area, average pore diameters and pore volumes obtained from air-dried and calcined particles. The surface area of the air-dried particles 119 m^2/g whereas, the particles are able to hold a higher surface area of 244 m^2/g after calcination at 800 °C (Table 5.1). This increase in the surface area may be attributed to the formation of the γ-Al_2O_3 phase upon calcination (Chuah, Jaenicke, & Xu, 2000). The surface area results of the particles reported here are comparable with previously published surface area data of alginate-boehmite particles (Prouzet et al., 2006).

The pore-size distribution curves for both the air-dried and calcined samples, calculated by Barret–Joyner–Halenda (BJH) method using nitrogen desorption branch are presented in Fig. 4.15. This is clearly seen in the corresponding pore size distribution plots, that the pore-size distributions are unimodal since one peak is observed in the mesopore region (2–50 nm). The peaks are centered at 9.2 and 10.5 nm in diameter for the air-dried (Fig. 4.15a) and calcined particles (Fig. 4.15b), respectively. The pore structure data obtained for both the air-dried and calcined sample are summarized in Table 5.1. As can be seen from Table 5.1, in case of air-

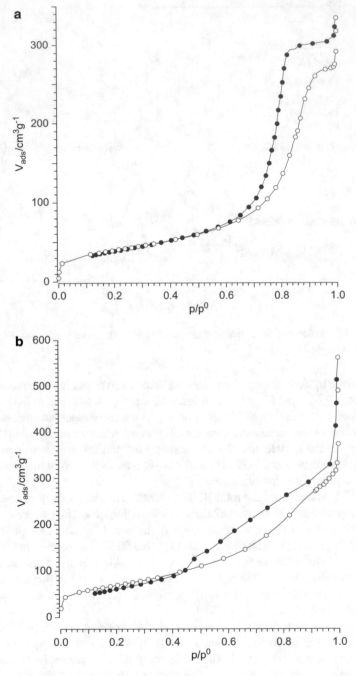

Fig. 4.14 Nitrogen adsorption–desorption isotherms for (**a**) air-dried and (**b**) calcined (800 °C) particles

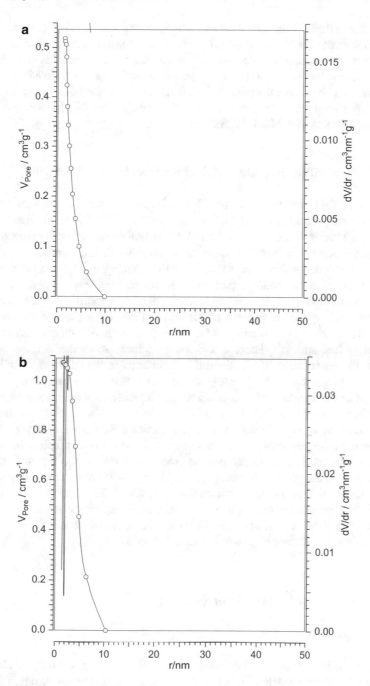

Fig. 4.15 Pore size distribution of the (**a**) air-dried and (**b**) calcined particles

dried sample, the average pore diameter was 8.40 nm while it was increased slightly to 10.60 nm after calcination at 800 °C. It is reasonable to assume that any alteration in mean pore volume after calcinations is attributed to removal of alginate from the particles as evident from XRD analysis (see curve c in Fig. 4.13). In addition, as an effect of the calcination, the pores might originate from the void spaces within the alumina particles due to the removal of water from the crystal planes of the aluminum oxide (Johnson & Mooi, 1968).

4.3.1.5   Mechanical Strength and Surface Morphology of Particles

The mechanical strength and morphological investigation of air-dried and calcined particles were performed. From the data reported in Table 5.1, it is evident that the value of mechanical strength of the air-dried particles was higher than those for the calcined particles. The higher mechanical strength of air-dried particles could be a result of the contribution at molecular level of binding of alginate gel combined with the cohesion of boehmite particles. However, calcining the particles caused alginate matrix to be totally removed from the particle network, as evidenced by the XRD analysis reported in curve c of Fig. 4.13. Thus, the alginate matrixes do not contribute anymore to the binding of the calcined particles which may have reduced the mechanical strength. There is evidence of increase in the volume of pores through the observation of N_2 desorption isotherms in the calcined particles (see Table 5.1). In general, lower defect (less porous) structure of a particle provides the benefit of higher mechanical strength. Thus, an increased pore volume may have led to a higher number of defects and this is likely to reduce the mechanical strength.

The morphological investigation of the particles was conducted via scanning electron microscope (SEM) as shown in Fig. 4.16. One distinct feature in the cross section of the particles after calcination (see Fig. 4.16b) is the increase in defect with respect to the porous channelling morphology with interconnectivity of pores in comparison to the air-dried particles (see Fig. 4.16a). These porous structures are of particular interest for the application to catalysis, especially, for solid-catalyzed reactions in petrochemicals, because of their unrestricted diffusion of reactants and reaction products (Linssen, Cassiers, Cool, & Vansant, 2003).

4.3.2   Oil Drop Granulation Process

4.3.2.1   Shape of Particles

To obtain the information on the change of the gel properties, viscosity, surface tension and density were determined as a function of boehmite concentration from 13 to 28 (%w/v) as shown in Table 4.4.

As can be seen from Table 4.4 that increasing the boehmite concentration in the sol leads to an increase of both density and viscosity, while decrease in surface tension. The visual observations of the droplet seen in Fig. 4.17 (I) to (VI) reveal that

Fig. 4.16 Scanning electron micrograph of the particles showing the cross-sectioned surface morphology, (**a**) air-dried particles, and (**b**) calcined particles

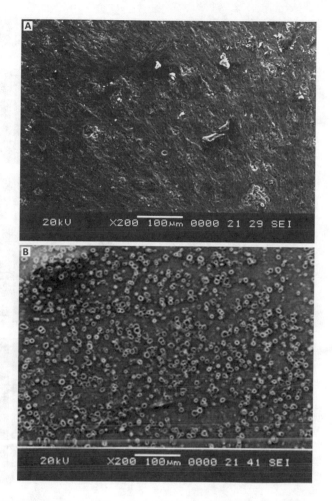

Table 4.4 Properties of the boehmite gel

Sample no.	Concentration (%w/v)	Viscosity η (mPa.s)	Density ρ (kg/m³)	Surface tension γ (mN/m)
S1	13	280.02	1070	66.15
S2	16	293.17	1088	63.14
S3	19	306.73	1107	59.10
S4	22	315.15	1125	57.30
S5	25	327.12	1144	52.07
S6	28	343.21	1161	50.89

droplet shape changes from a spherical to a tailing body, as the concentration of boehmite increases. In fact, the extent to which droplets change their shape can be described by their gel properties, written as viscosity, density and surface tension. In the progression of droplet formation implicates three major scenarios, i.e.,

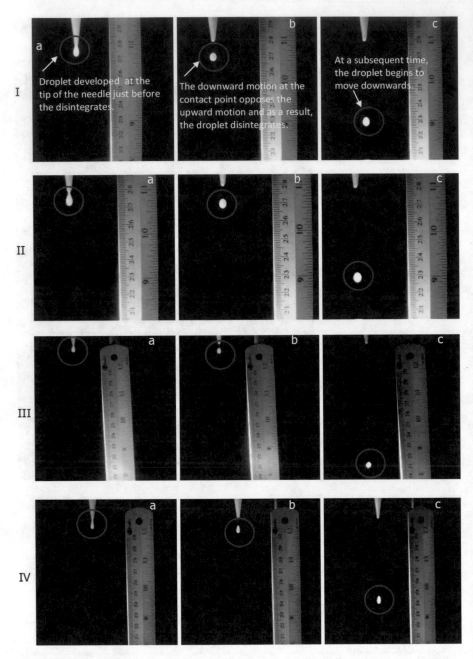

Fig. 4.17 Photographs of the progression of droplet in air phase (**a**) undetached drop holding at the tip of the needle; (**b**) droplets just after disintegration; and (**c**) further progression of droplet towards the downward direction. The photograph obtained with various initial boehmite concentrations; I. 13 % (w/v), II. 16 % (w/v), III. 19 % (w/v), IV. 22 % (w/v), V. 25 % (w/v), VI. 28 % (w/v)

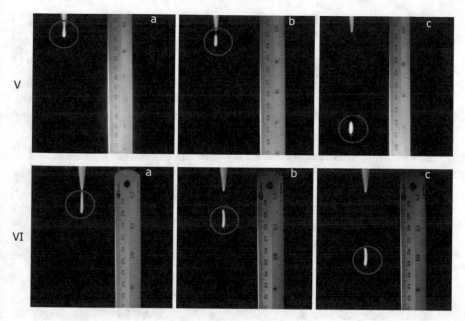

Fig. 4.17 (continued)

expansion of the gel protrusion, breakup of the gel ligament, and completion of the main/satellite droplet generation as shown in Fig. 4.17a–c.

As revealed in Fig. 4.17 (I) to (II), the shape of the droplet remains close to spherical in shape throughout the progression of droplet formation process. This behavior occurs in droplets with relatively low viscosity (i.e., higher viscous forces) and low density and also for droplets with higher surface tension (see Table 4.4). A somewhat related conclusion has been reached by Taylor (1966), and observed that the liquid droplets (water/oil) that have reached the liquid–air interface could be retained their spherical form which may be results of only under specific combination of physical properties of the liquid (viscosity, surface tension) at a fixed liquid flow rate. Other process variable including the aging time of particles in the ammonia solution could also be involved on the formation of spherical alumina particle (Buelna & Lin, 1999).

As the concentration of boehmite increases, the behavior of the ejected droplet changes to a slender liquid ligament as seen in Fig. 4.17 (III) to (VI). It seemed reasonable to point out that a smaller value of surface tension represents a weaker cohesive force, which leads to a slender liquid ligament with a relatively longer breakup length (see Table 4.4).

Subsequently, that the droplet was kept under continuous observation while passing through a oil layer shown in Fig. 4.18 (I) to (VI). It is apparent from the visualized images of the droplet that there is no detectable change in the shape of drop as the drop continues its downward motion. A combination of the effect of gel properties may lead to the observed trends in Fig. 4.19.

For purposes of comparison, the particles sphericity factor is plotted in Fig. 4.20 as a function of boehmite at different concentrations (%w/v). The results represent an

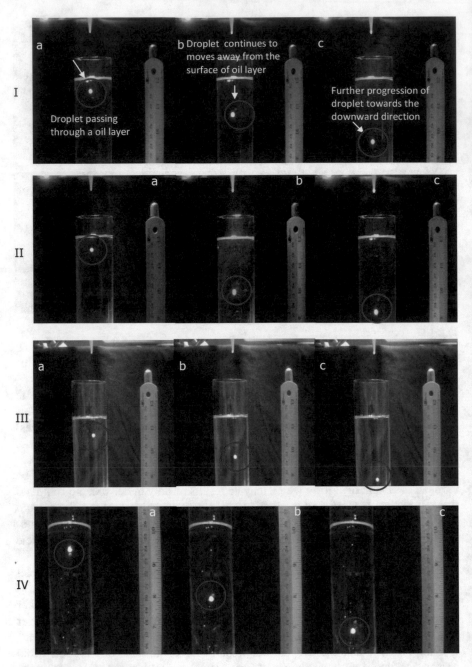

Fig. 4.18 Photographs of the progression of droplet in paraffin oil layer phase; (**a**) near to the surface of oil layer, (**b**) away from the surface of oil layer, and (**c**) further continues to move in the downward direction. The photograph obtained with various initial boehmite concentrations (g/L); I. 13 % (w/v), II. 16 % (w/v), III. 19 % (w/v), IV. 22 % (w/v), V. 25 % (w/v), VI. 28 % (w/v)

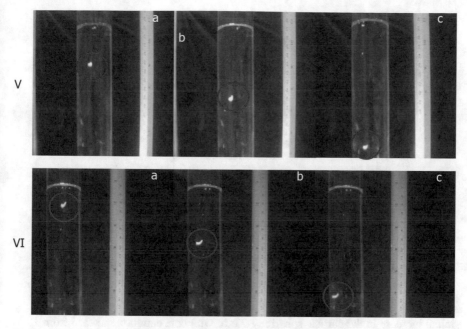

V

VI

Fig. 4.18 (continued)

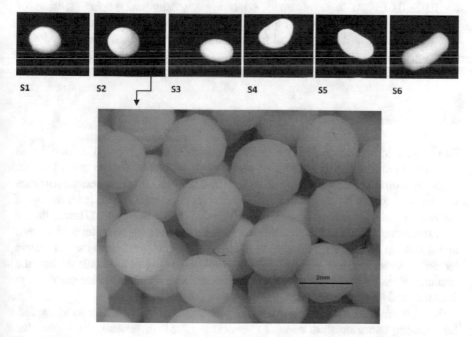

S1   S2   S3   S4   S5   S6

Fig. 4.19 Optical photograph shows the effect of boehmite concentration on sphericity factor of the particle

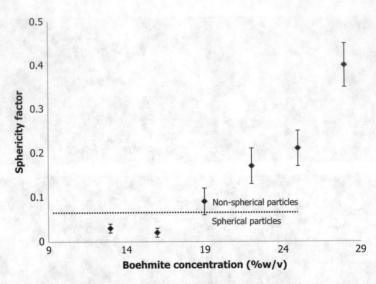

Fig. 4.20 Effect of boehmite concentration on sphericity factor of the particles

initial increase in the particle sphericity factor with concentration increase from 13 to 16% (w/v) and then, the change in the sphericity factor is significant as the concentration increases from a value of 19 to 28 %w/v. The particles with lowest sphericity factor 0.02 ± 0.01 (S2) were obtained exhibiting viscosity and surface tension of 293.17 mPa.s and 63.14 mN/m respectively. It has been found that a particle can be considered spherical if the spherical factor is less than 0.05 (Chan, Lee, Ravindra, & Poncelet, 2009). It can be stated from the foregoing work and those of the previous works, the gel properties have to be accounted for the formation of spherical particles.

4.3.2.2   Size of Particles

To obtain information on the size of the particles, the size prediction model was therefore used to measure the size of particles. It should be kept in mind that the particles with a sphericity factor of 0.02 ± 0.01 (S2) named as spherical particles will be used for the future discussion. The droplet size while falling from the tip of the needle had an average diameter of 4.26 mm as shown in Fig. 4.21. Immediately after the gelation for 1 h, the particle contracted to an average diameter of 3.96 mm and shrinkage of 7%, might be due to the expulsion of solvent from the gel matrix termed as syneresis, as reported by Dijk and Schenk (1984). Upon air drying, the particles shrunk substantially and their diameter and shrinkage were changed to 2.23 mm and 48%, respectively.

In most cases, the degree of shrinkage varies considerably owing to many factors, among them moisture content (Fennema, 1975). In general, the higher the moisture content, the higher the release of the moisture, and thus, reduction the volume of the particle is higher.

Droplets while falling into gelling bath (d= 4.26 mm in diameter)

size reduced by 7%

Particles after gelling for 1 hours ($d_p = 3.96$ mm in diameter, k=0.85)

size reduced by 48%

Particles after drying for 12 hours ($d_p = 2.23$ mm in diameter, k=0.48)

size reduced by 54%

Particles after calcination at 800 °C for 3 hours ($d_p = 1.96$ mm in diameter, k=0.42)

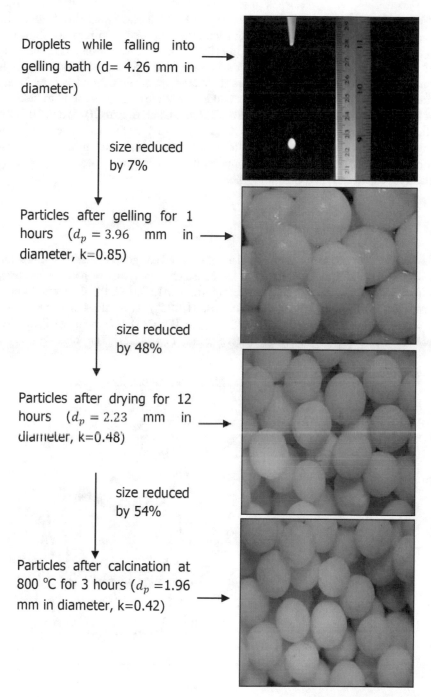

Fig. 4.21 Changes in particle size after each preparation steps (Boehmite concentration=16% (w/v) (for spherical particles)

When the particles were calcined at up to 800 °C, it underwent shrinkage to the extent of 54% and the corresponding diameter was then 1.96 mm, as shown in Fig. 4.21. As reported by Deng and Lin (1997) that the calcination causes the sol–gel particles to shrink and harden to a certain extent as the additional water is removed from the particle. The volume shrinkage upon calcination has been investigated for a number of particles types and temperatures and was found to be 50 % for Ru/γ-Al$_2$O$_3$ granular particles at 600 °C (Kim, Choi, & Kim 2005) and 70 % for Silicalite/γ-Alumina Granules at 450 °C (Yang & Lin, 2000). The observed variations could also be attributed to the differences in the interfacial tension and viscosity of fluids and the dependence of temperature on these properties, as reported by Siladitya, Chatterjee, and Ganguli (1999).

4.3.2.3 Structure of Particles

The XRD studies, carried out on powders of crushed sol-gel prepared alumina granules, provided information relating to the phase characteristics of particle, as structural changes took place during the calcinations. At 300 °C, the diffraction peaks corresponding to Al$_2$O$_3$ JCPDS File No. 00-031-0026) at 2θ values of 19.58, 32.77, 37.60, 39.49, 44.53 and carbon (JCPDS File No. 00-026-1077) owing to diffraction peak at 66.76 of 2θ appeared (curve a in Fig. 4.22). The reason for the

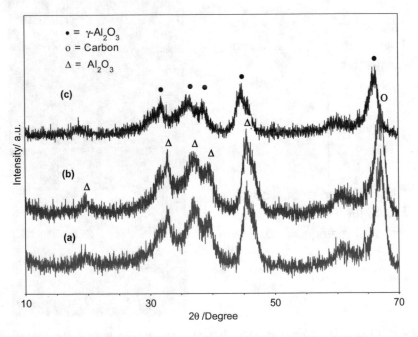

Fig. 4.22 XRD pattern of the spherical particles after calcination for 3 h at various temperatures: (**a**) 300 °C, (**b**) 500 °C, (**c**) 800 °C

transformation to Al_2O_3 from $AlO(OH)$ was associated with the dehydration of the sample occurring at high temperatures (>200 °C) (Vazquez et al., 1997). As far as could be determined by X-ray diffraction, no new crystalline phase was observed as the calcination temperature increases from 300 to 500 °C (curve b in Fig. 4.22). When the temperature increased to 800 °C, the carbon phase disappeared and only the gamma alumina (γ-Al_2O_3) phase (JCPDS File No. 00-029-0063) appeared 31.93, 37.60, 39.49, 45.79 and 66.73 of 2θ due to the occurrence of the crystalline phase transformation (curve c in Fig. 4.22). The absence of a carbon peak at 2θ values of at 66.76 of also confirms that there is no carbon phase in the material. Previous work has confirmed that the alumina phases are transformed accompanied with the increased calcination temperature (Levin & Brandon, 1998) and the γ-Al_2O_3 phase of the sample may be caused by the higher calcination temperature.

4.3.2.4 Surface Area and Porosity of Particles

The nitrogen adsorption–desorption isotherms of the uncalcined and calcined particles are shown in Fig. 4.23; their porosity and surface area are also listed in Table 4.5. All the isotherms are of type IV, characteristic of mesoporous materials, according to the IUPAC classification. The isotherms for both of calcined and uncalcined particles present H1-type hysteresis loops characteristic (the two branches remain nearly vertical and parallel over an appreciable range of gas uptake), this type is often associated with mesoporous or macroporous materials that have cylindrical pore structures (Sharma, Kumar, Saxena, Chand, & Gupta, 2002, Wahab & Ha, 2005).

The major uptake observed at high relative pressures (P/P_0) range 0.42-0.85 for air-dried particles (Fig. 4.23a) and at 0.70-0.86 for the calcined particles (Fig. 4.23b). Moreover, hysteresis loops with relatively sharp steep desorption were observed in the calcined particles (Fig. 4.23b) at high relative pressure (P/P_0) range probably due to adsorbate condensation in the larger mesopores (Hu & Srinivasan, 1999).

The pore-size distribution curves of the air-dried and calcined particles, determined by the BJH (Barrett, Joyner, and Halenda) method from the adsorption branch of the isotherms. The curves exhibit monomodal distribution (Fig. 4.24) with one centered in the mesoporous range (2–50 nm in diameter), indicating the uniform pore distribution within the particles. As can be seen from Table 6.2, the calcinated particle gives higher pore volume of 1.09 cm^3g^{-1} as compared to 0.46 cm^3g^{-1} in the case of air-dried particles. An observation somewhat analogous to that reported for the case of aluminum oxide by Johnson and Mooi (1968) and suggested that a significant amount of pores could be originated from the spaces within the alumina particles at high temperature (>700 °C) due to the removal of water from the crystal planes of the aluminum oxide. In addition, it might be reasonable to state that any alteration in pore volume after calcinations could be attributed to removal of starch from the particles as evident from XRD analysis cited above (see curve c in Fig. 4.22). The BET surface area is 178 m^2/g and 349 m^2/g for the air-dried and the calcined particle respectively.

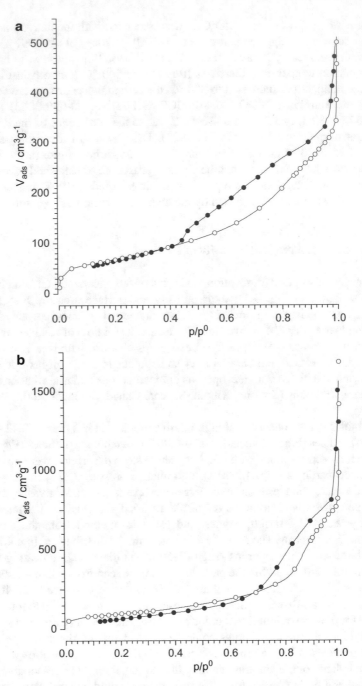

Fig. 4.23 N$_2$ adsorption–desorption isotherms of (**a**) air-dried particles and (**b**) calcined particles

Table 4.5 The porous and mechanical properties of the air-dried and calcinated particles

Parameters	Air-dried particle (2.23 mm)	Calcined particle (1.96 mm)
Pore volume (cm³/g)	0.46	1.09
Pore diameter (nm)	9.7	11.2
BET surface area (m²/g)	178	349
Crush strength (N/particle)	45 ± 3	51 ± 4

It is also constructive to compare the surface area of the synthesized γ-Al_2O_3 particles in this study with those of γ-Al_2O_3 particles synthesized by other researchers. The comparison of the surface area tabulated in Table 4.6 reveals that surface area the synthesized γ-Al_2O_3 particles is comparable to those reported in the literature, even though the calcination temperature in this study was slightly more severe.

4.3.2.5   Mechanical Strength and Surface Morphology of Particles

As shown in Table 4.5, the crush strength before calcination was 45 ± 3 N/particle, where the crush strength slightly increased to 51 ± 4 N/particle after calcined at 800 °C. According to Buelna, Lin, Liu, and Litster (2003), the enhancement of mechanical strength of the particle synthesized by sol-gel process are attributed to the formation of solid network with highly interconnected particles during the calcination which apparently facilitates highly packed structure. The shrinkage of the particles after calcinations may lead to increase the mechanical strength of particles, as evident from image analysis (Fig. 4.21). A somewhat analogues observation has been made by Buelna and Lin (1999) and reported that the gel shrunks under capillary force during the calcination as the liquid evaporates and thus, increases the mechanical strength of the particles.

To investigate further the morphological characteristics of particles, the SEM analysis was performed. As evidenced by Fig. 4.25, SEM images of the calcinated particles at different magnifications exhibited a porous channelling morphology with interconnectivity of pores (Fig. 4.25c, d) in comparison to the air-dried particles which were characterized with a smooth surface (Fig. 4.25a, b). The more porous structure of the particles may be beneficial in terms of enhancing the diffusion of the relatively large reactant and product molecules in the particles (Cejka & Mintova, 2007). This is also a significant point to be underscored in this study.

4.4   Comparison of Particle Properties

The properties of particle produced from integrated gelling process and oil drop granulation process was compared. It can be seen from the Table 4.7 that the surface area of the particle is 349 m²/g which is higher than that of particle derived from integrated gelling process (244 m²/g). The higher surface area could be beneficial to

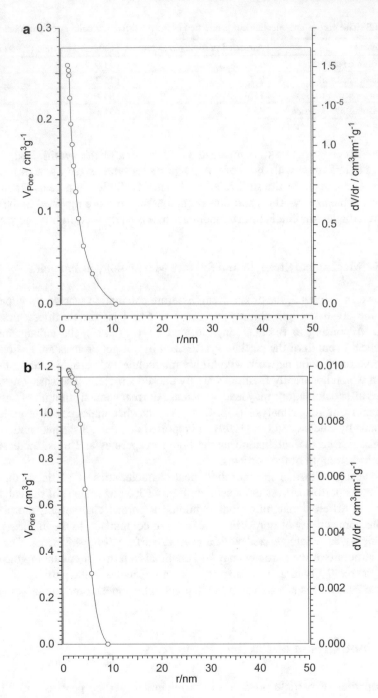

Fig. 4.24 BJH pore size distribution curves of (**a**) air-dried and (**b**) calcined particles

Table 4.6 Comparison of surface area of the particles

[a]Particle	Surface area (m²/g)	Particle size (mm)	Calcination temperature (°C)	Reference
CuO/Al_2O_3	393	2	450	Yang and Lin (2000)
$SiO2/\gamma\text{-}Al_2O_3$	336	2	450	Wang and Lin (1998)
$\gamma\text{-}Al_2O_3$	349	2	800	This study

[a]Particles were prepared using oil drop granulation process

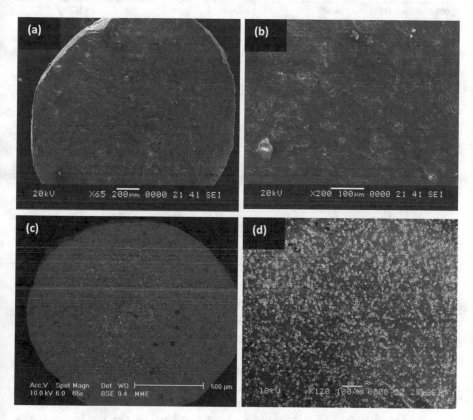

Fig. 4.25 Scanning electron micrograph of the particles showing the cross-sectioned surface morphology at different magnifications; (**a**, **b**) air-dried particles, and (**c**, **d**) calcined particles

provide the large accessible for the catalyst to react. In addition, it was found that the particles produced by the oil-drop granulation process had higher mechanical strength which could be crucial for long term stability of particle in reaction environment.

In integrated gelling process, alginate was used as gelling agent for boehmite particles. Thus, the binding of alginate at a molecular level with boehmite particle

Table 4.7 Comparison of particle properties

Parameters	Integrated gelling process		Oil drop granulation process	
	Air-dried particles	Calcined particles	Air-dried particles	Calcined particles
Sphericity factor (SF)	0.03±0.02	0.03±0.02	0.02±0.01	0.02±0.01
Size of the particle (mm)	2.17	≈2	2.12	≈2
Structure by XRD	AlOOH	γ-Al$_2$O$_3$	AlOOH	γ-Al$_2$O$_3$
BET Surface area (m^2/g)	119	244	178	349
pore diameter (nm)	8.40	10.60	9.7	11.2
Pore volume (cm^3/g)	0.44	0.83	0.46	1.09
Mechanical strength (N/particle)±S.D.	5±1.0	3±0.4	45±3	51±4
Morphology by SEM	Nonporous	Porous	Nonporous	Porous

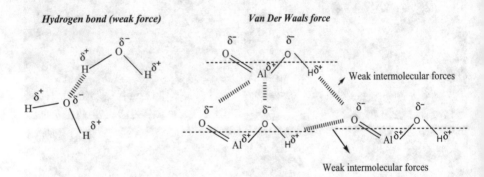

Fig. 4.26 Weak forces of attraction in the boehmite gelling process

could be associated with the mechanical strength of particle. However, the alginate matrix was removed from the particle network after calcination, as evidenced by the XRD analysis (curve c of Fig. 4.13). Thus, the alginate matrixes do not contribute anymore to the binding of the calcined particles which may have reduced the mechanical strength. On the other hand, in the oil drop granulation process, the boehmite colloidal suspension was transform into a gel at lower pH (pH 1) might be due to the weak forces of attraction i.e., hydrogen bond or van der Waals forces (Fig. 4.26). During the formation of gel, water may be trapped in the gel. Therefore, the hydrogen bond between particles may strengthen their interparticle attraction forces. In addition, van der Waals forces between particle may contribute to increase the interparticle attraction forces under acidic condition. According to Buelna et al. (2003), the enhancement of mechanical strength of the particle synthesized by oil drop granulation process was attributed to the formation of solid network with highly interconnected particles during the calcination which apparently facilitates highly packed structure. Therefore, the particle obtained from oil drop granulation process possessed higher mechanical strength compared to the integrated gelling process.

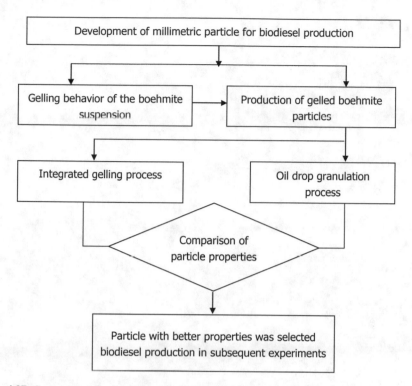

Fig. 4.27 Summary of the development of millimetric particle for biodiesel production

Besides, the lowest SF was obtained from Oil drop granulation process which could minimize the abrasion of catalyst in the reaction medium. Therefore, the particle derived from oil-drop granulation process was selected as a catalyst support for the production of biodiesel in subsequent experiments.

4.5 Summary

The development of millimetric particle for biodiesel production in this chapter is divided into three parts. In the first part, the study of the gelling behavior of the boehmite suspension, followed by production of gelled boehmite particles using two different approaches: integrated gelling process and Oil drop granulation process. Finally, the properties between two approaches were compared and better ones were selected for biodiesel production in subsequent experiments. The summary is provided in the following flowchart (Fig. 4.27).

Chapter 5
Production of Biodiesel Using Spherical Millimetric Catalyst

5.1 Introduction

This chapter demonstrates the use of oil drop granulation process derived spherical millimetric gamma alumina (γ-Al_2O_3) particle supported by the potassium iodide (KI), potassium fluoride (KF), and sodium nitrate ($NaNO_3$) catalysts for the production of biodiesel from palm oil at 60 °C in batch process. The catalysts were prepared at different loading contents, and the effects of these catalysts on the transesterification reaction were evaluated in terms of the FAMEs yield. The characteristics of the synthesized catalysts, such as the structural, textural, and base properties were carried out using X-ray diffraction (XRD), N_2 adsorption–desorption isotherms, and CO_2-temperature-programmed reduction (TPD) techniques, respectively. The relationship between the basicity of the catalyst and their catalytic activity in the transesterification of palm oil was also discussed. Finally, the reusability of the single catalyst was determined and the leachate of catalyst into the reaction product was verified by X-ray fluorescence (XRF).

5.2 Structure of Catalyst

Figure 5.1 shows the XRD patterns of the KI/γ-Al_2O_3 catalyst calcined at 500 °C. . In the sample with a low loading of $0.06g_{KI}/g_{\gamma-Al2O3}$ (curve a in Fig. 5.1), the characteristic peaks appear at $2\theta = 31.9°$, 37.6°, 39.5°, 45.8°, and 66.7° are assigned to γ-Alumina (JCPDS File No. 00-029-0063) and the characteristics peak of KI peak could not be detected. The absence of KI suggests that KI is well dispersed on the support at 0.06 $g_{KI}/g_{\gamma-Al2O3}$ as a monolayer (Jiang et al., 2001). When the catalyst loading was increased to $0.15g_{KI}/g_{\gamma-Al2O3}$ (curve b in Fig. 5.1), the characteristic peaks of KI (JCPDS File No. 00-004-0471) ($2\theta = 21.7°$, 25.2°) can be observed although the intensities of these peaks are not as high as those observed for the γ-alumina ($2\theta = 31.9°$, 37.6°, 39.5°, 45.8°, and 66.7°).

© Springer International Publishing Switzerland 2017
A. Islam, P. Ravindra, *Biodiesel Production with Green Technologies*,
DOI 10.1007/978-3-319-45273-9_5

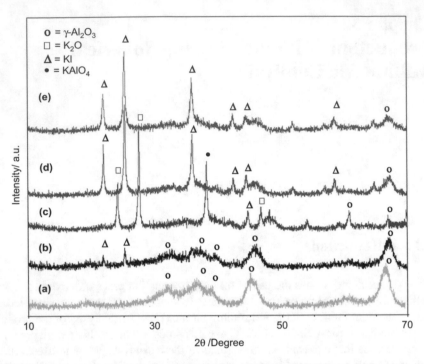

Fig. 5.1 XRD spectra of γ-Al₂O₃ particles with different catalyst loading (**a**) 0.06 g $g_{KI}/g_{γAl2O3}$, (**b**) $0.15g_{KI}/g_{γAl2O3}$, (**c**) $0.24g_{KI}/g_{γAl2O3}$, (**d**) $0.30g_{KI}/g_{γAl2O3}$, (**e**) $0.33g_{KI}/g_{γAl2O3}$

An additional K_2O phase ($2\theta = 24.5°$, $27.1°$, $45.1°$) and potassium aluminum oxide phase ($2\theta = 37.9°$) appear when the catalyst loading was increased to $0.24g_{KI}/g_{γ-Al2O3}$ (curves c in Fig. 5.1). It implies that KI interacts with the support and formed new phases. A similar observation was obtained in the case of KNO_3/Al_2O_3 where it was found that the alkali metal oxides and alkali-aluminate crystalline structure were the catalytically active sites to catalyze the transesterification reactions (Xie, Peng, & Chen, 2006b). In addition, the peaks of γ-alumina ($2\theta = 60.5°$, $66.7°$) and KI ($2\theta = 44.3°$) phases were observed at the loading of $0.24g_{KI}/g_{γ-Al2O3}$. When the loading amount achieved $0.30g_{KI}/g_{γ-Al2O3}$, the characteristics peaks of γ-alumina (JCPDS File No. 00-029-0063) at $2\theta = 66.7°$ and potassium iodide (JCPDS File No. 00-004-0471) at $2\theta = 21.7°$, $25.2°$, $35.9°$, $42.4°$, $44.3°$, and $58.3°$ was appeared clearly in the diffraction patterns (curve d in Fig. 5.1). As far as could be determined by XRD, no new phase transformation was took place when the KI loading was raised to $0.33g_{KI}/g_{γ-Al2O3}$ (curve e in Fig. 5.1).

The XRD patterns of the KF/γ-Al₂O₃ catalyst calcined at 500 °C are presented in Fig. 5.2. As for the KF content of 0.06 g/g γ-Al₂O₃ (curve a), the characteristics peaks of γ-Al₂O₃ (JCPDS File No. 00-029-0063) at $2\theta = 45.79°$ and $66.76°$, and KF (JCPDS File No. 00-036-1458) at $2\theta = 28.89°$ were observed. The intensity of the peaks associated with the KF increased as the loading amount was increased to 0.15 g/g γ-Al₂O₃ (curve b). New diffraction lines (curve c) assigned to K_2O (JCPDS

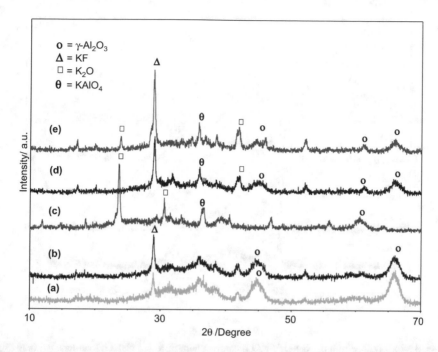

Fig. 5.2 XRD spectrum of KF/γ-Al$_2$O$_3$ catalyst with different KF loadings (**a**) 0.06 g/g γ-Al$_2$O$_3$, (**b**) 0.15 g/g γ-Al$_2$O$_3$, (**c**) 0.24 g/g γ-Al$_2$O$_3$, (**d**) 0.30 g/g γ-Al$_2$O$_3$, (**e**) 0.33 g/g γ-Al$_2$O$_3$

File No. 00-023-0493) at $2\theta = 23.88°$ and 30.0°, and KAlF$_4$ (JCPDS File No. 00-040-0549) at $2\theta = 35.52°$ appeared with a further increase in the KF catalyst loading to 0.24 g/g γ-Al$_2$O$_3$. The undetectable phase of KF at this loading may have decomposed, interacted with the γ-Al$_2$O$_3$ and formed new phases of K$_2$O and KAlF$_4$. At 0.30 g/g γ-Al$_2$O$_3$ (curve d), the KF diffraction line appeared at $2\theta = 28.89°$ and was more intense when the KF loading was increased to 0.33 g/γ-Al$_2$O$_3$ (curve e).

Figure 5.3 shows the XRD patterns of the NaNO$_3$/γ-Al$_2$O$_3$ catalyst calcined at 500 °C. Only diffraction lines of γ-Al$_2$O$_3$ (JCPDS File No. 00-029-0063) at $2\theta = 45.79°$ and 66.76° were visible in the case of 0.06 g/g γ-Al$_2$O$_3$ (curve a). At this loading, the NaNO$_3$ phase could not be detected, which may have been due to the high dispersion of NaNO$_3$ on the γ-Al$_2$O$_3$ support in the form of a monolayer (Jiang et al., 2001). When the loading amount was increased to 0.15 g/g γ-Al$_2$O$_3$, the formation of NaAlO$_2$ (JCPDS File No. 00-019-1179) at $2\theta = 30.38°$ and Na$_2$O (JCPDS File No. 00-023-0528) at $2\theta = 32.17°$, along with the γ-Al$_2$O$_3$ phases, were noted (curve b). As the loading amount was increased to 0.24 and 0.30 g/g γ-Al$_2$O$_3$, the relative intensity of the peak corresponding to the NaAlO$_2$ phase was found to increase (curve c), indicating an increase in the degree of crystallinity of the phase. When the NaNO$_3$ content was increased to 0.33 g/γ-Al$_2$O$_3$, a strong diffraction peak attributed to NaNO$_3$ (JCPDS File No. 00-006-0392) phase appeared at 29.69°, implying that a residual bulk phase of NaNO$_3$ remained in the γ-Al$_2$O$_3$ support (curve e). It is worth noting that the reflection peaks assigned to NaNO$_3$ were

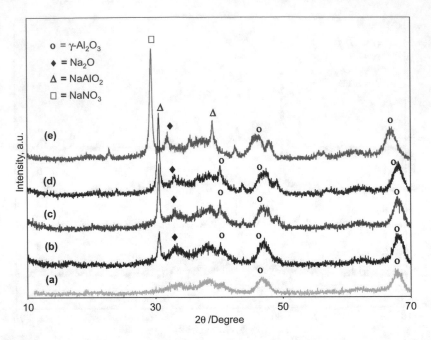

Fig. 5.3 XRD spectrum of NaNO₃/γ-Al₂O₃ catalyst with different NaNO₃ loadings: (**a**) 0.06 g/g γ-Al₂O₃, (**b**) 0.15 g/g γ-Al₂O₃, (**c**) 0.24 g/g γ-Al₂O₃, (**d**) 0.30 g/g γ-Al₂O₃, (**e**) 0.33 g/g γ-Al₂O₃

not observed when the $NaNO_3$ content was below 0.33 g/γ-Al₂O₃, indicating decomposition of the $NaNO_3$ on the γ-Al₂O₃ support.

5.3 Properties of Catalyst

5.3.1 Basicity of Catalyst

The strength and number of basic sites of the catalysts were analyzed by CO_2-TPD. In general, the stronger the basic site of a catalyst, the higher the temperature is needed to desorb the adsorbed CO_2. It has been reported that the basicity of supported basic catalyst is dependent on the presence of oxide or/and hydroxide groups on the catalyst surface (Cantrell, Gillie, Lee, & Wilson, 2005; Meher, Kulkarni, Dalai, & Naik, 2006a; Serio et al., 2006).

The TPD profiles of desorbed CO_2 on KI/γ-Al₂O₃, KF/γ-Al₂O₃, and NaNO₃/γ--Al₂O₃ catalysts are shown in Figs. 5.4, 5.5, and 5.6, respectively. The results showed complex desorption profiles, which could be due to the presence of a variety of basic sites of different strength. As reported by researchers that the desorption peaks at temperatures <200 °C, 200–400 °C and >400 °C correspond to the weak, medium and strong basic strength, respectively (Sun et al., 2007; Yin, Xiao, & Song, 2008;

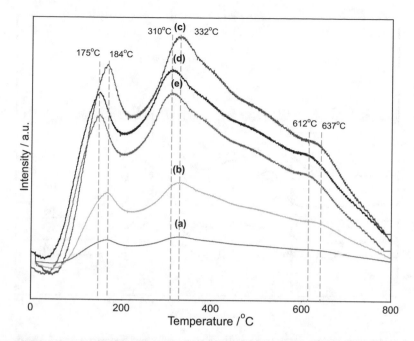

Fig. 5.4 CO_2-TPD spectra of γ-Al_2O_3 particles with different catalyst loading (**a**) 0.06 $g_{KI}/g_{\gamma Al2O3}$, (**b**) 0.15$g_{KI}/g_{\gamma Al2O3}$, (**c**) 0.24$g_{KI}/g_{\gamma Al2O3}$, (**d**) 0.30$g_{KI}/g_{\gamma Al2O3}$, (**e**) 0.33$g_{KI}/g_{\gamma Al2O3}$

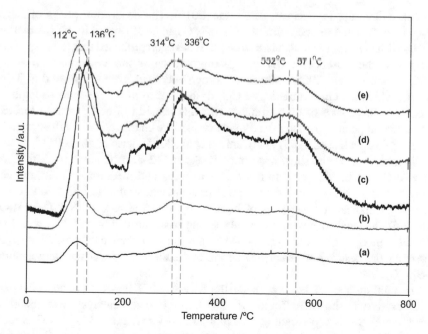

Fig. 5.5 CO_2-TPD profiles of KF/γ-Al_2O_3 catalyst with different KF loadings: (**a**) 0.06 g/g γ-Al_2O_3, (**b**) 0.15 g/g γ-Al_2O_3, (**c**) 0.24 g/g γ-Al_2O_3, (**d**) 0.30 g/g γ-Al_2O_3, (**e**) 0.33 g/g γAl_2O_3

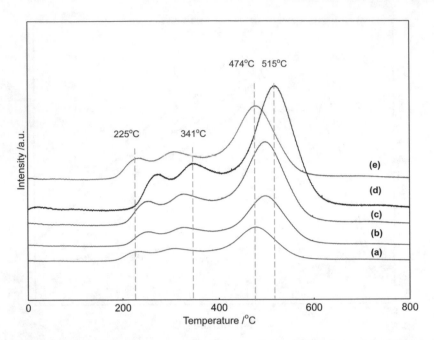

Fig. 5.6 CO_2-TPD profiles of $NaNO_3/\gamma$-Al_2O_3 catalyst with different $NaNO_3$ loadings: (**a**) 0.06 g/g γ-Al_2O_3, (**b**) 0.15 g/g γ-Al_2O_3, (**c**) 0.24 g/g γ-Al_2O_3, (**d**) 0.30 g/g γ-Al_2O_3, (**e**) 0.33 g/g γ-Al_2O_3

Yoosuk, Udomsap, Puttasawat, & Krasae, 2010). The desorption peaks of KI/γ-Al_2O_3 (Fig. 5.4) at a temperature of 175–184 °C can be attributed to the interaction of CO_2 with the sites of weak basic strength. It has been proposed that these sites are assigned to the presence of OH^- groups on the catalyst surface (Bolognini et al., 2002; Cantrell et al., 2005). Moreover, strong and sharp peaks appeared at 310–332 °C, which are characteristics of CO_2 desorption from the medium strength of basic sites. Basic sites of high strength, which is related to free O^{2-} anions located in a particular position of the mixed oxide surface are characterized by the presence of desorption peak at 612–637 °C (Cantrell et al., 2005; Serio et al., 2006).

Similarly, the results show desorption peaks at 112–136 °C and 314–336 °C of KF/γ-Al_2O_3 can be attributed to basic sites of weak and medium strength, respectively (Fig. 5.5). Moreover, the desorption peaks at temperatures of 552–571 °C for KF/γ-Al_2O_3 indicated basic sites of high strength. In the case of $NaNO_3/\gamma$-Al_2O_3 (Fig. 5.6), the catalyst possesses mainly strong basic sites (474–515 °C) with low amount of moderate basic sites (225–341 °C). Among the three catalysts, KI/γ-Al_2O_3 possesses the desorption peak with strong basic sites at a very high temperature (>600 °C).

The total number of basic sites for the KI/γ-Al_2O_3, KF/γ-Al_2O_3, and $NaNO_3/\gamma$-Al_2O_3 catalyst is shown in Table 5.1. The basicity of the catalysts was found to increase with the catalyst loading. The highest basicity was found to be 867 μmol/g catalyst, 703 μmol/g catalyst, and 760 μmol/g catalyst at a KI content of 0.24 g/g γ-Al_2O_3, KF content of 0.24 g/g γ-Al_2O_3 and $NaNO_3$ content of 0.30 g/g γ-Al_2O_3,

Table 5.1 CO$_2$-TPD spectrum of millimetric γ-Al$_2$O$_3$ particles with different catalyst loadings

Catalyst loading (g/g γ-Al$_2$O$_3$)	KI/γ-Al$_2$O$_3$ CO$_2$ desorbed (μmol/g catalyst)	KF/γ-Al$_2$O$_3$ CO$_2$ desorbed (μmol/g catalyst)	NaNO$_3$/γ-Al$_2$O$_3$ CO$_2$ desorbed (μmol/g catalyst)
0.06	198	304	408
0.15	318	505	509
0.24	867	703	690
0.30	466	611	760
0.33	457	604	700

respectively. The results showed that the basic sites as well as the basic strength of the impregnated catalysts were affected by the catalyst loading, which was in good agreement with the findings of other authors (Meher, Sagar, & Naik, 2006; Verziu et al., 2009). A mechanism has been proposed by Xie, Peng, and Chen (2006a, 2006b); Xie et al. (2006b) to explain the decrease of basic sites with increasing catalyst loading.

It has been suggested that the vacancies formed in the alumina structure during the thermal treatment favors the migration of metal ions of salt to the vacant sites that subsequently decompose to form basic sites. When the amount of salt loaded on γ-Al$_2$O$_3$ is more than the saturation uptake, it could not likely be decomposed to create basic sites on the catalyst. It is evident from the TPD profile that the KI/γ-Al$_2$O$_3$ catalyst (Fig. 5.4) showed more basic sites with a higher intensity of strongly basic sites compared with the KF/γ-Al$_2$O$_3$ (Fig. 5.5) and NaNO$_3$/γ-Al$_2$O$_3$ catalyst (Fig. 5.6).

5.3.2   Surface Area and Pore Structure of Catalyst

Figure 5.7 shows the nitrogen adsorption–desorption isotherm of the γ-Al$_2$O$_3$ catalysts support. As shown in Fig. 7.7, the isotherm of the γ-Al$_2$O$_3$ support can be described as a type IV isotherm according to the IUPAC classifications. The isotherm displayed a steep hysteresis loop at relative pressures (P/P$_0$) in the range between 0.75 and 0.97, and it showed a similar shape to those reported for mesoporous materials (Melero et al., 2002). The pore size distributions of the catalyst are unimodal as can be seen in Fig. 5.8. The unimodal pore size distribution indicated the presence of monodisperse pores in the particles. Table 5.2 compiles the surface area, pore volume and pore diameter of the catalysts.

As shown in Table 5.2, the γ-Al$_2$O$_3$ support gave a high surface area of 349 m^2/g, indicating a high thermal stability of the γ-Al$_2$O$_3$ support (Cao et al., 2008). In the case of the KI/γ-Al$_2$O$_3$ catalyst the surface area decreased from 339 to 133 m^2/g, while the pore diameter decreased from 9 to 3.5 nm (Table 5.2). It is clearly observed that with an increase in catalyst content from 0.06 g to 0.33 g/g γ-Al$_2$O$_3$, the surface area of the catalyst decreased (Table 5.2), along with a decrease in the pore volume and pore size. Similar trends were observed with the other two catalysts, namely

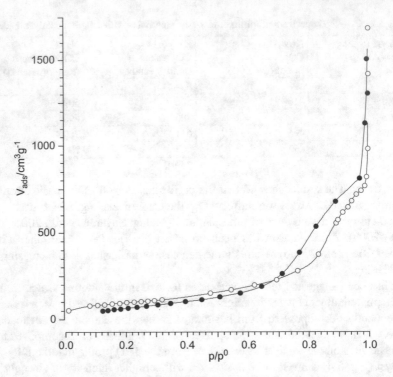

Fig. 5.7 N₂ adsorption–desorption isotherm of the millimetric γ-Al₂O₃ particles

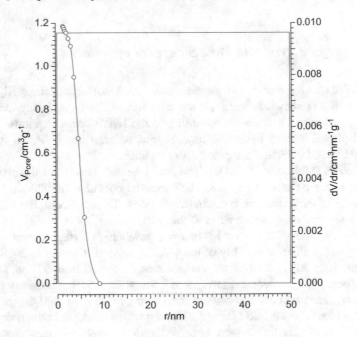

Fig. 5.8 BJH pore size distribution (b) of the millimetric γ-Al₂O₃ particles

Table 5.2 BET surface area and pore structure of millimetric γ-Al$_2$O$_3$ particles with different catalyst loadings

Catalyst type	Catalyst loading (g/g γ-Al$_2$O$_3$)	Pore volume (cm^3/g)	Surface area (m^2/g)	Pore diameter (nm)
γ-Al$_2$O$_3$	–	1.38	349	11.2
KI/γ-Al$_2$O$_3$	0.06	1.09	326	9.0
	0.15	0.84	299	7.3
	0.24	0.63	277	7
	0.30	0.49	244	3.9
	0.33	0.43	133	3.5
KF/γ-Al$_2$O$_3$	0.06	1.21	319	9.2
	0.15	0.92	291	7.5
	0.24	0.85	265	6.9
	0.30	0.65	199	5.7
	0.33	0.53	143	3.9
NaNO$_3$/γ-Al$_2$O$_3$	0.06	1.19	322	9.4
	0.15	0.97	304	7.3
	0.24	0.83	287	6.2
	0.30	0.65	203	5.1
	0.33	0.49	152	4.1

KF/γ-Al$_2$O$_3$ and NaNO$_3$/γ-Al$_2$O$_3$. This gradual loss of surface area has been previously observed and attributed to the effect of catalyst deposition on the support, which results in partial blocking of the porous network (Boz et al., 2009; Meher et al., 2006).

5.4 Biodiesel Production

The activity of catalysts was evaluated in the transesterification of palm oil with methanol in terms of fatty acid methyl ester (FAME) yield at different catalyst loadings, as shown in Fig. 5.9. The FAME yield was found to increase with catalyst loading where the highest yield of 98 % for KI/γ-Al$_2$O$_3$ and 80 % for KF/γ-Al$_2$O$_3$ catalyst were obtained at a loading of 0.24 g/g γ-Al$_2$O$_3$. However, further increase in the catalyst loading was found to reduce the FAME yield. On the other hand, the highest FAME yield obtained for the NaNO$_3$/γ-Al$_2$O$_3$ catalyst was 87 % at a loading of 0.3 g /g γ-Al$_2$O$_3$.

The variation in the FAME yield could be attributed to the new phases generated with the increase in catalyst loading. As shown in Figs. 5.2 and 5.3, the high yield of biodiesel can be correlated to the generation of Na$_2$O and NaAlO$_2$ phases for the NaNO$_3$/γ-Al$_2$O$_3$ catalyst, and K$_2$O and KAlF$_4$ phases for the KF/γ-Al$_2$O$_3$ catalyst. Similarly, the XRD pattern of the KI/γ-Al$_2$O$_3$ catalyst (see curve c in Fig. 5.1) reveals that the K$_2$O phase and metal aluminate phase formed at catalyst loading of

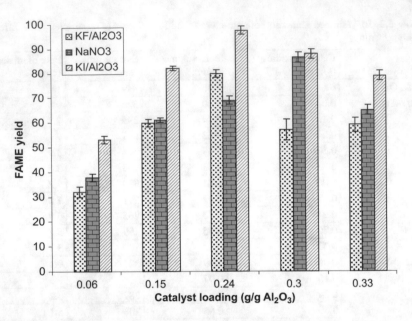

Fig. 5.9 Effect of KI and NaNO₃ catalyst loadings on FAME yield. Reaction conditions: methanol–oil molar ratio of 14:1, catalyst amount 0.6 g (4 wt%, $g_{cat.}/g_{oil}$, calcined at 500 °C), reaction temperature of 60 °C, reaction time of 4 h

$0.24_{gKI}/g_{\gamma\text{-}Al2O3}$ can be the considered as the active sites for the transesterification reactions, as also proposed by Xie et al. (2006a). Moreover, the biodiesel yield could also be affected by a difference in the number of basic sites as well as the level of strongly basic sites present in the catalysts. The higher level of basic sites and strength of the (curve c in Fig. 5.4) KI/γ-Al₂O₃ catalysts might result in a higher biodiesel yield compared with the KF/γ-Al₂O₃ (Fig. 5.5) and KF/γ-Al₂O₃ catalyst (Fig. 5.6). Similar phenomenon has been observed in previous work on KF/MgO catalyst (Wan et al., 2008). Although all the impregnated catalysts exhibited a decrease in surface area, pore diameter and pore volume at higher loadings, their pore structures were found to be that of mesopores (i.e., >2 nm in diameter) (Table 5.2). It has been proposed that mesoporous catalysts permit the diffusion of large molecules like triglycerides into the active sites (Macario et al., 2010). Therefore, mesoporous characteristics of the catalysts may have an influence on FAME yield.

A comparison of similar catalysts from previous studies is compiled in Table 5.3. A direct comparison of the biodiesel yield between the studies is difficult because of the variation in the reaction conditions, notably, the oil to methanol molar ratio, reaction time and amount of catalyst used. In addition, the reported results showed that the structure and composition of supported catalysts may affect both the FAME yield as well as its productivity. However, the biodiesel yield produced using the spherical millimetric γ-Al₂O₃ supported catalyst was comparable to that produced using catalysts of smaller size.

Table 5.3 Comparison of biodiesel yields

Catalyst	Reaction condition							Calcination		[g]BY (%)	[h]BP	Reuse catalyst		References
	Oil type	[a]RT (°C)	[b]Cat. (g)	oil (g)	[c]M/O	[d]Rt (h)	[e]BD (g)	[f]CT (°C)	t (h)			cycle	[g]BY (%)	
KI/Al$_2$O$_3$	Soybean	MR	0.4	16	15:1	8	15.36	500	3	96	0.12	–	–	Xie et al., 2006a
KOH/Al$_2$O$_3$	Palm	60	3	100	15:1	2	91	500	3	91	0.45	–	–	Noiroj et al., 2009
Ca(NO$_3$)$_2$/Al$_2$O$_3$	Palm	60	1	10	65:1	3	9.4	450	4	94	0.31	–	–	Benjag. et al., 2009
KNO$_3$/Al$_2$O$_3$	Palm	60	1	10	65:1	3	9.47	550	4	94.7	0.31	–	–	Benjag, et al. 2009
KF/Al$_2$O$_3$	Sunflower	75	0.25	28	4:1	2	27.44	450	2	98	0.47	4	88	Verziu et al., 2009
Egg shell	Soybean	65	0.69	23	9:1	3	21.85	800	2	95.5	0.32	13	95	Wei et al., 2009
CaO/Al$_2$O$_3$	Palm	65	3	50	12:1	5	49.32	718	5	98.64	0.19	2	91	Zabeti et al., 2010
KOH/AC	Palm	64	5	50	24:1	1	49	–	–	98	0.98	3	87	Baroutian et al., 2010
C$_4$H$_4$O$_6$HK/ZrO$_2$	Soybean	60	1	18.22	16:1	2	17.86	512	5	98.03	0.49	5	89	Qiu et al., 2011
KI/γAl$_2$O$_3$	Palm	60	0.6	15	14:1	4	14.5	500	3	98	0.24	11	79	Present study
[e]KF/Al$_2$O$_3$	Sunflower	72	4.6	23	4:1	2	12.65	450	2	55	0.27	–	–	Verziu et al., 2009

(continued)

Table 5.3 (continued)

Catalyst	Reaction condition							Calcination				Reuse catalyst		References
	Oil type	[a]RT (°C)	[b]Cat. (g)	oil (g)	[c]M/O	[d]Rt (h)	[e]BD (g)	[f]CT (°C)	t (h)	[g]BY (%)	[h]BP	cycle	[g]BY (%)	
KF/nano-γ-Al$_2$O$_3$	Canola oil	65	7.5	50	15:1	8	48.5	500	3	97	0.12	–	–	Boz et al., 2009
NaNO$_3$/Al$_2$O$_3$	Palm	65	1	10	65:1	3	9.3	350	2	93	0.31	–	–	Benjap et al., 2006

All of the reactions were performed under atmospheric pressure

[a]Reaction temperature

[b]Catalyst used for reaction

[c]Molar ratio of methanol to oil

[d]Reaction time

[e]Biodiesel

[f]Calcination temperature of the catalyst

[g]Biodiesel yield (%) = biodiesel(g)/oil (g) × 100

[h]Biodiesel productivity = biodiesel (g)/oil(g).time

[i]Activated carbon

5.5   Optimization of Process Variables and Reusability of KI/γ-Al₂O₃ Catalyst

Since the preliminary experiments showed that KI/γAl₂O₃ gave the highest FAME yield, the catalyst process parameter using the catalyst was optimized and the catalyst reusability were conducted.

5.5.1   Effect of Reaction Time on the FAME Yield

Reaction time plays a crucial in biodiesel production as it can influence the FAME yield and economic consideration. The effect of reaction time on the FAME yield was investigated, as shown in Fig. 5.10. The FAME yield increased from 11 to 98 % as the reaction time was increased from 1 to 4 h and remained constant thereafter, indicating an equilibrium FAME yield.

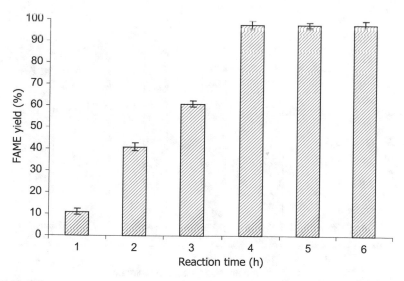

Fig. 5.10 Effect of reaction time on FAME yield. Reaction conditions: methanol–oil molar ratio 14:1, catalyst loading $0.24 g_{KI}/g_{\gamma Al2O3}$, reaction temperature 60 °C, catalyst amount 0.6 g (4 wt%, $g_{Cat.}/g_{oil}$), and catalyst calcined at 500 °C

5.5.2   Effect of Oil and Methanol Ratio on the FAME Yield

Figure 5.11 shows the effect of oil and methanol molar ratio on the FAME yield. According to reaction stoichiometry, three moles of methanol are required to react with each mole of triglyceride, but in practice a higher molar ratio is employed in order to drive the reaction towards completion and produce more methyl esters as product. The yield was found to increase considerably and reached a maximum when the methanol–oil molar ratio was increased to 14, beyond which no further increment in yield could be observed. The observations are in agreement with previous reports (Liu et al., 2007).

5.5.3   Catalyst Reusability

The stability of KI/γ-Al$_2$O$_3$ catalyst was evaluated by performing the reusability test on the catalyst. Figure 5.12 presents the effect of catalyst reuse on the FAME yield. The amount of the biodiesel yields after the 11th cycle obtained in this work was 79%, which is comparable in biodiesel yield to that of the results obtained in previous studies (see Table 5.3).

The XRF results (Table 5.4) show that the amount of K$_2$O in the catalyst, after the first and eleventh cycles, was 53% and 26%, respectively. This could be due to the leaching of K$_2$O from the γ-Al$_2$O$_3$ support. Although the catalyst was reused for 11 times, it showed good mechanical stability as the support remained spherical with no

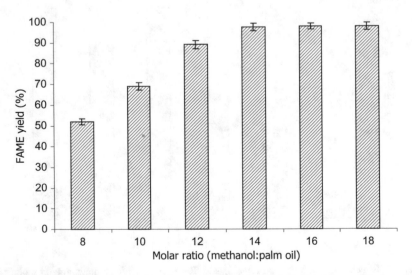

Fig. 5.11 Influence of methanol–oil molar ratio on the FAME yield. reaction conditions: catalyst loading, 0.24g_{KI}/$g_{\gamma\text{-Al2O3}}$, reaction temperature 60 °C, reaction time 4 h, catalyst amount 0.6 g (4 wt%, $g_{cat.}$/g_{oil}), catalyst calcined at 500 °C

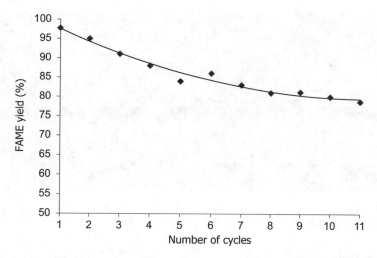

Fig. 5.12 Reusability of KI/γ-Al$_2$O$_3$ catalyst in transesterification of palm oil. Reaction conditions: methanol–oil molar ratio 14:1, catalyst loading 0.24g_{KI}/$g_{\gamma Al2O3}$, reaction temperature 60 °C, reaction time 4 h, catalyst amount 0.6 g (4 wt%, $g_{Cat.}$/g_{oil}), and catalyst calcined at 500 °C

Table 5.4 Composition of reused KI/γ-Al$_2$O$_3$

	Elemental composition (%)		
Component	1st reuse	7th reuse	11th reuse
Al$_2$O$_3$	45.964	67.162	73.63
K$_2$O	53.319	32.296	26.14
SO$_3$	0.58	0.435	0.16
CuO	0.087	0.067	0.01
ZnO	0.05	0.04	0.03

Reaction conditions: methanol–oil molar ratio 14:1, catalyst loading 0.24g_{KI}/$g_{\gamma Al2O3}$, reaction temperature 60 °C, reaction time 4 h, catalyst amount 0.6 g (4 wt%, $g_{Cat.}$/g_{oil}), and catalyst calcined at 500 °C

sign of disintegration (see Fig. 5.13). This indicates its potential for use in industrial environment. The optimized parameter for biodiesel production of KI/γ-Al$_2$O$_3$ catalyst was given in Table 5.5.

The use of spherical millimetric γ-Al$_2$O$_3$ support particles loaded with KI, KF and NaNO$_3$ catalysts for the transesterification of palm oil and methanol to biodiesel was demonstrated. The highest FAME yield obtained from KI/γ-Al$_2$O$_3$ catalyst was 98 % after 4 h of reaction time at 60 °C and the yield was found to directly correspond to the catalyst basicity. It can be correlated with their generation of K$_2$O, KAlO$_2$ for KI/γ-Al$_2$O$_3$ catalyst as evident from XRD which were possibly the main active sites for the transesterification reaction. Similarly, the activity of KF/Al$_2$O$_3$ catalysts was remarkably improved when the catalysts loading were 0.30 g (g_{Cat}/$g_{\gamma-Al2O3}$) for NaNO$_3$/γ-Al$_2$O$_3$ and 0.24 g (g_{Cat}/$g_{\gamma-Al2O3}$) for KF/γ-Al$_2$O$_3$. The number

Fig. 5.13 Optical photograph of KI/γ-Al$_2$O$_3$ catalyst (**a**) before and (**b**) after use in transesterification of palm oil with methanol

Table 5.5 Optimized parameter for biodiesel production of KI/γ-Al$_2$O$_3$ catalyst

Process parameter	Optimized parameter
Catalyst loading	0.24g$_{KI}$/g$_{γ-Al2O3}$
Methanol–oil ratio	14:1
Reaction time	4 h
Reuse	79% yield (11 cycle)

Reaction conditions: reaction temperature 60 °C, catalyst amount 0.6 g (4 wt%, g$_{cat.}$/g$_{oil}$), catalyst calcined at 500 °C

and strength of basic sites in the catalysts could be attributed to its catalytic activity in transesterification, as evidenced by CO$_2$-TPD. The high activity towards the transesterification reaction corresponds to the generation of Na$_2$O, NaAlO$_2$ on NaNO$_3$/γ-Al$_2$O$_3$ catalyst, K$_2$O, KAlF$_4$ on KF/γ-Al$_2$O$_3$ catalyst. The mesoporous characteristics of the catalysts may have an influence on FAME yield as evident from N$_2$ adsorption–desorption isotherm, since the triglycerides molecules could diffuse into the mosepores of the catalyst. The KI/γ-Al$_2$O$_3$ catalyst exhibited good operational stability with biodiesel yield of 79% even after 11 cycles of successive reuse. The activity of reused KI/γ-Al$_2$O$_3$ catalyst is proportional to the K$_2$O concentration in the catalyst sample, **as revealed** from **XRF** results. In fact, this works have explored the potential **applicability** of the millimetric catalyst for transesterification of palm oil with methanol.

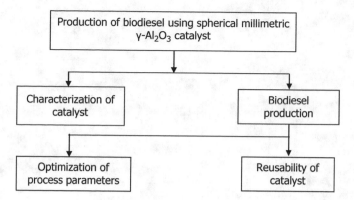

Fig. 5.14 Summary of the biodiesel production using millimetric catalyst

5.6   Summary

The production of biodiesel using spherical millimetric γ-Al_2O_3 catalyst is summarized in the following flowchart (Fig. 5.14).

Chapter 6
Conclusions

In conclusion, from the rheological study, boehmite suspensions under controlled pH conditions show a non-Newtonian behavior that could result from particle–particle aggregation due to weak attractive forces. The correlations found were that the consistency index (k) increased with decreasing suspension pH and in turn increased suspension viscosity. The flow behavior index (n) increases with increasing pH, reaching a maximum value at pH 7.6. The results show that the flow activation energy (ΔE) response of boehmite suspensions is sensitive to a change in pH. Similar to the flow behavior index, decreasing the pH to 1 resulted in an increase in the ΔE value, suggesting the rigidity of the gel network. Based on the results, the suspensions with the high pH showed high sensitivity to temperature with low activation energy. The correlation between the rheology of the suspension and the density of the boehmite sol suggests that increased interparticle interactions at low pH may be the cause of the higher density.

A spherical millimetric catalyst particle for biodiesel production has been developed by using two different synthetic approaches, i.e., (a) integrated method and (b) sol–gel method. In integrated gelling process, the $CaCl_2$ concentration (g/L) in the gelling bath containing 36.5 g/L $AlCl_3$ is a critical factor for controlling the shape of beads. The spherical beads (sphericity factor=0.03±0.01) was obtained at the volume fractions of 2.5 g/L $CaCl_2$, followed by calcination at 800 °C afforded a γ-Al_2O_3 phase with porous network of the beads. On the other hand, in determining the shape evolutions of the droplet as a function of boehmite concentration using sol–gel process, it was observed that the ejected liquid with relatively high surface tensions and low viscosity generally brings about a spherical droplet. The beads with lowest sphericity factor 0.02±0.01 were obtained exhibiting viscosity and surface tension of 293.17 mPa.s and 63.14 mN/m, respectively.

The synthesized beads were dried in the air, followed by calcination at 800 °C, where the crystalline structure of γ-Al_2O_3 phase was formed. The N_2 adsorption–desorption isotherm and BJH pore size distribution of the synthesized bead elucidate that the existence of mesoporous characteristic with very narrow monomodal pore in the mesopore region. The highest surface area (349 m^2g^{-1}) and pore volume

© Springer International Publishing Switzerland 2017
A. Islam, P. Ravindra, *Biodiesel Production with Green Technologies*,
DOI 10.1007/978-3-319-45273-9_6

$(1.09 \text{ cm}^3\text{g}^{-1})$ were obtained from sol–gel process. SEM images of the cross sectioned beads exhibited a porous channelling morphology with interconnectivity of pores which might be beneficial in terms of enhancing the diffusion of the relatively large reactant and product molecules. Bead made from sol–gel process had higher mechanical strength, which was crucial for long term stability of catalyst. Therefore, the millimetric spherical bead derived from sol–gel process was used as a catalyst support in this study for transesterification reaction at different reaction conditions.

A spherical millimetric catalyst particle has been used for transesterification of palm oil with methanol. The catalytic activity was controlled by optimizing the reaction time, methanol–palm oil molar ratio, catalyst–support loading ratio, catalyst amount, and calcination temperature. The highest FAME yield obtained from $KI/\gamma\text{-Al}_2\text{O}_3$ catalyst was 98 % after 4 h of reaction time at 60 °C and the yield was found to directly correspond to the catalyst basicity. It can be correlated with their generation of $K_2\text{O}$, $KAlO_2$ for $KI/\gamma\text{-Al}_2\text{O}_3$ catalyst as evident from XRD which were possibly the main active sites for the transesterification reaction. Similarly, the activity of $KF/\text{Al}_2\text{O}_3$ catalysts was remarkably improved when the catalysts loading were 0.30 g ($g_{Cat}/g_{\gamma\text{-Al2O3}}$) for $NaNO_3/\gamma\text{-Al}_2\text{O}_3$ and 0.24 g ($g_{Cat}/g_{\gamma\text{-Al2O3}}$) for $KF/\gamma\text{-Al}_2\text{O}_3$. The number and strength of basic sites in the catalysts could be attributed to its catalytic activity in transesterification, as evidenced by CO_2-TPD. The generation of $Na_2\text{O}$, $NaAlO_2$ on $NaNO_3/\gamma\text{-Al}_2\text{O}_3$ catalyst, $K_2\text{O}$, $KAlF_4$ on $KF/\gamma\text{-Al}_2\text{O}_3$ catalyst which were possibly the active sites for the transesterification reaction. The high FAME yield could also be attributed to the mesoporous characteristic of the catalyst with pore diameter of 7–9 nm as evident from N_2 adsorption–desorption isotherm, since the smaller triglycerides molecules could diffuse into the catalyst.

The $KI/\gamma\text{-Al}_2\text{O}_3$ catalyst exhibited good operational stability with biodiesel yield of 79 % even after 11 cycles of successive reuse. The activity of reused $KI/\gamma\text{-Al}_2\text{O}_3$ catalyst is proportional to the $K_2\text{O}$ concentration in the catalyst sample, **as revealed** from **XRF** results. In fact, this works have explored the potential **applicability** of the millimetric catalyst for transesterification of palm oil with methanol. Based on this research, further investigations should focus on the following respects: (a) further study that needs to be addressed is the improvement of the chemical stability of the millimetric catalysts in order to allow regeneration and reuse without loss of activity; (b) to explain the actual reaction mechanism, further research on the kinetics of the reaction that take place on the millimetric catalyst surface will be required, so that rate of the transesterification reaction can be further improved; (c) impregnation efficiency and loss of solid materials from catalyst and support during calcinations.

References

Abdullah, A. Z., Razali, N., & Lee, K. T. (2009). Optimization of mesoporous K/SBA-15 catalyzed transesterification of palm oil using response surface methodology. *Fuel Processing Technology, 90*, 958–964.

Abreu, F. R., Alves, M. B., Macedo, C. C. S., Zara, L. F., & Suarez, P. A. Z. (2005). New multiphase catalytic systems based on tin compounds active for vegetable oil transesterification reaction. *Journal of Molecular Catalysis A: Chemical, 227*, 263–267.

Abreu, F. R., Lima, D. G., Hamu, E. H., Einloft, S., Rubim, J. C., & Suarez, P. A. Z. (2003). New metal catalysts for soybean oil transesterification. *Journal of the American Chemical Society, 80*, 601–604.

Aderemi, B. O., & Hameed, B. H. (2009). Alum as a heterogeneous catalyst for the transesterification of palm oil. *Applied Catalysis A: General, 370*(1), 54–58.

Akoh, C. C., Chang, S. W., Lee, G. C., Shaw, J. F., Akoh, C. C., Chang, S. W., et al. (2007). Enzymatic Approach to Biodiesel Production. *Journal of Agricultural and Food Chemistry, 55*, 8995–9005.

Albuquerque, M. C. G., Gonzalez, J. S., Robles, J. M. M., Tost, R. M., Castellon, E. R., Lopez, A. J., et al. (2008). MgM (M = Al and Ca) oxides as basic catalysts in transesterification processes. *Applied Catalysis A: General, 347*, 162–168.

Almeida, R. M. D., Noda, L. K., Goncalves, N. S., Meneghetti, S. M. P., & Meneghetti, M. R. (2008). Transesterification reaction of vegetable oils, using superacid sulfated TiO₂–base catalysts. *Applied Catalysis A: General, 347*, 100–105.

Alonso, R. D., Mariscal, M., Tost, R. M., Poves, M. D. Z., & Granados, M. L. (2007). Potassium leaching during triglyceride transesterification using K/γ-Al₂O₃ catalysts. *Catalysis Communications, 8*, 2074–2080.

Amigun, B., Sigamoney, R., & Blottnitz, H. V. (2008). Commercialisation of biofuel industry in Africa: A review. *Renewable and Sustainable Energy Reviews, 12*, 690–711.

Antunes, W. M., Veloso, C. D. O., Assumpc, C., & Henriques, O. (2008). Transesterification of soybean oil with methanol catalyzed by basic solids. *Catalysis Today, 133–135*, 548–554.

Araujoa, L. R. R., Scofielda, C. F., Pasturaa, N. M. R., & Gonzalezb, W. A. (2006). H₃PO₄/Al₂O₃ catalysts: Characterization and catalytic evaluation of oleic acid conversion to biofuels and biolubricant. *Materials Research, 9*, 181–184.

Ardizzone, S., Bianchi, C. L., Ragaini, V., & Vercelli, B. (1999). SO₄–ZrO₂ catalysts for the esterification of benzoic acid to methylbenzoate. *Catalysis Letters, 62*, 59–65.

Arzamendi, G., Campoa, I., Arguinarena, E., Sanchez, M., Montes, M., & Gandıa, L. M. (2007). Synthesis of biodiesel with heterogeneous NaOH/alumina catalysts: Comparison with homogeneous NaOH. *Chemical Engineering Journal, 134*, 123–130.

© Springer International Publishing Switzerland 2017
A. Islam, P. Ravindra, *Biodiesel Production with Green Technologies*,
DOI 10.1007/978-3-319-45273-9

Backov, R. (2006). Combining soft matter and soft chemistry: integrative chemistry towards designing novel and complex multiscale architectures. *Soft Materials, 2*, 452–464.

Baronetti, G., Padro, C., Scelza, O., & Castro, A. (1993). Structure and reactivity of alkali doped calcium oxide catalysts for oxidative coupling of methane. *Applied Catalysis A: General, 101*, 167–183.

Baroutian, S., Aroua, M. K., Abdul Raman, A. A., & Sulaiman, N. M. N. (2010). Potassium hydroxide catalyst supported on palm shell activated carbon for transesterification of palm oil. *Fuel Processing Technology, 91*, 1378–1385.

Beck, J. S., Vartuli, J. C., Roth, W. J., Leonowicz, M. E., Kresge, C. T., Schmitt, K. D., et al. (1992). A New Family of Mesoporous Molecular Sieves Prepared with Liquid Crystal Templates. *Journal of the American Chemical Society, 114*, 10834–10843.

Benjag S., Ngamcharussrivichai, C., Bunyakiat, K. (2009). Al2O3- supported alkali and alkali earth metal oxides for transesterification of palm kernel oil and coconut oil. *Chemical Engineering Journal, 145*, 468–474.

Benjapornkulaphong, S., Ngamcharussrivichai, C., & Bunyakiat, K. (2009). Al_2O_3-supported alkali and alkali earth metal oxides for transesterification of palm kernel oil and coconut oil. *Chemical Engineering Journal, 145*, 468–474.

Bernardes, O. L., Bevilaqua, J. V., Leal, M. C. M. R., Freire, D. M. G., & Langone, M. A. P. (2007). Biodiesel fuel production by the transesterification reaction of soybean oil using immobilized lipase. *Applied Biochemistry and Biotechnology, 137–140*, 105–114.

Bertgeret, G., & Gallezot, P. (1997). *Handbook of heterogeneous catalysis* (pp. 439–464). Weinheim: Wiley-VCH.

Bo, X., Guomin, X., Lingfeng, C., Ruiping, W., & Lijing, G. (2007). Transesterification of Palm Oil with Methanol to Biodiesel over a KF/Al_2O_3 Heterogeneous Base Catalyst. *Energy & Fuels, 21*, 3109–3112.

Bolognini, M., Cavani, F., Scagliarini, D., Flego, C., Perego, C., & Saba, M. (2002). Heterogeneous basic catalysts as alternatives to homogeneous catalysts: reactivity of Mg/Al mixed oxides in the alkylation of m-cresol with methanol. *Catalysis Today, 75*, 103–111.

Bols, M., & Skrydstrup, T. (1995). Silicon-Tethered Reactions. *Chemistry Review, 95*, 1253–1277.

Bota, R. M., Houthoofd, K., Grobet, P. J., & Jacobs, P. A. (2010). Superbase catalysts from thermally decomposed sodium azide supported on mesoporous γ-alumina. *Catalysis Today, 152*, 99–103.

Bourikas, K., Kordulis, C., & Lycourghiotis, A. (2006). The Role of the Liquid-Solid Interface in the Preparation of Supported Catalysts. *Catalysis Reviews, 48*, 363–444.

Boyse, R. A., & Ko, E. I. (1999). Commercially available zirconia–tungstate as a benchmark catalytic material. *Applied Catalysis A: General, 177*, 131–137.

Boz, N., Degirmenbasi, N., & Kalyon, D. M. (2009). Conversion of biomass to fuel: Transesterification of vegetable oil to biodiesel using KF loaded nano-γ-Al_2O_3 as catalyst. *Applied Catalysis B: Environmental, 89*, 590–596.

Brito, A., Arvelo, R., Borges, M. E., Garcia, F., Garcia, M. T., Diaz, M. C., et al. (2007). Reuse of fried oil to obtain biodiesel: zeolite Y as catalyst. *International Journal of Chemical Reactor Engineering, 5*, 1–13.

Buchmeiser, M. R. (2001). New synthetic ways for the preparation of high-performance liquid chromatography supports. *Journal of Chromatography. A, 918*, 233–266.

Buelna, G., & Lin, Y. S. (1999). Sol–gel-derived mesoporous γ-alumina granules. *Microporous and Mesoporous Materials, 30*, 359–369.

Buelna, G., & Lin, Y. S. (2004). Characteristics and desulfurization-regeneration properties of sol–gel-derived copper oxide on alumina sorbents. *Separation and Purification Technology, 39*, 167–179.

Buelna, G., Lin, Y. S., Liu, L. X., & Litster, J. D. (2003). Structural and Mechanical Properties of Nanostructured Granular Alumina Catalysts. *Industrial and Engineering Chemistry Research, 42*, 442–447.

Bunyakiat, K., Makmee, S., Sawangkeaw, R., & Ngamprasertsith, S. (2006). Continuous Production of Biodiesel via Transesterification from Vegetable Oils in Supercritical Methanol. *Energy & Fuels, 20*, 812–817.

Burgess, J. C. (1990). The contribution of efficient energy pricing to reducing carbon dioxide emissions. *Energy Policy, 18*, 449–455.

Campanati, M., Fornasari, G., & Vaccari, A. (2003). A Fundamentals in the preparation of heterogeneous catalysts. *Catalysis Today, 77*, 299–314.

Canakci, M., & Gerpen, J. V. (1999). Biodiesel Production via Acid Catalysis. *Transactions of ASAE, 42*, 1203–1210.

Cantrell, D. G., Gillie, L. J., Lee, A. F., & Wilson, K. (2005). Structure–reactivity correlations in MgAl hydrotalcite catalysts for biodiesel synthesis. *Applied Catalysis A: General, 287*, 183–190.

Cao, J. L., Wang, Y., Zhang, T. Y., Wu, S. H., & Yuan, Z. Y. (2008). Preparation, characterization and catalytic behavior of nanostructured mesoporous $CuO/CeO.8ZrO_2O_2$ catalysts for low-temperature CO oxidation. *Applied Catalysis B Environmental, 78*, 120–128.

Carmo, A. C., Jr., de Souza, L. K. C., da Costa, C. E. F., Longo, E., Zamian, J. R., & da Rocha Filho, G. N. (2009). Production of biodiesel by esterification of palmitic acid over Mesoporous aluminosilicate Al-MCM-41. *Fuel., 88*, 461–468.

Cejka, J., & Mintova, S. (2007). Perpectives of micro/mesoporous composites in catalysis. *Catalysis Reviews, 49*, 457–509.

Centi, G., & Perathoner, S. (2003a). Integrated design for solid catalysts in multiphase reactions. *CATTECH, 7*, 78–89.

Centi, G., & Perathoner, S. (2003b). Novel catalyst design for multiphase reactions. *Catalysis Today, 3*, 79–80.

Cervero, J. M., Coca, J., & Luque, S. (2008). Production of biodiesel from vegetable oils. *Grasas Y Aceites. International Journal of Fats and Oils, 59*, 76–83.

Chan, E. S. (2011). Preparation of Ca-alginate beads containing high oil content: Influence of process variables on encapsulation efficiency and bead properties. *Carbohydrate Polymers, 88*, 1267–1275.

Chan, E. S., Lee, B. B., Ravindra, P., & Poncelet, D. (2009). Shape and size analysis of ca alginate particles produced through extrusion-dripping method. *Journal of Colloid and Interface Science, 338*, 63–72.

Cheng, L. J., Lan, X., Feng, X., Zhan-Wen, W., & Fei, W. (2006). Effect of hydrothermal treatment on the acidity distribution of γ-Al_2O_3 support. *Applied Surface Science, 253*, 766–770.

Chhinnan, M. S., Mcwatters, K. H., & Rao, V. N. M. (1985). Rheological Characterization of Grain Legume Pastes and Effect of Hydration Time and Water Level on Apparent Viscosity. *Journal of Food Science, 50*, 1167–1171.

Chorkendorff, I., & Niemantsverdriet, J. W. (2003). *Concepts of Modern Catalysis and Kinetics.* Germany: Wiley-VCH.

Chuah, G. K., Jaenicke, S., & Xu, T. H. (2000). The effect of digestion on the surface area and porosity of alumina. *Microporous and Mesoporous Materials, 37*, 345–353.

Cole-Hamilton, D. J. (2003). Homogeneous catalysis-new approaches to catalyst separation, recovery, and recycling. *Science, 299*, 1702–1706.

Cordeiro, C. S., Arizaga, G. G. C., Ramos, L. P., & Wypych, F. (2008). A new zinc hydroxide nitrate heterogeneous catalyst for the esterification of free fatty acids and the transesterification of vegetable oils. *Catalysis Communications, 9*, 2140–2143.

Corma, A. (1995). Inorganic Solid Acids and Their Use in Acid-Catalyzed Hydrocarbon Reactions. *Chemistry Review, 95*, 559–614.

Corma, A. (1997). From Microporous to Mesoporous Molecular Sieve Materials and Their Use in Catalysis. *Chemistry Review, 97*, 2373–2420.

Corma, A., Diaz-Cabanas, M. J., Jorda, J. L., Martinez, C., & Moliner, M. (2006). High-throughput synthesis and catalytic properties of a molecular sieve with 18- and 10-member rings. *Nature, 443*, 842–845.

Cosimo, J. I. D., Diez, V. K., Xu, M., Iglesi, E., & Apestegui, C. R. (1998). Structure and Surface and Catalytic Properties of Mg-Al Basic Oxides. *Journal of Catalysis, 178*, 499–510.

Crabba, E., Nolasco-Hipolito, C., Kobayashi, G., Sonomoto, K., & Ishizaki, A. (2001). Biodiesel Production from Crude Palm Oil and Evaluation of Butanol Extraction and Fuel Properties. *Process Biochemistry, 37*, 67–71.

Cristiani, C., Valentini, M., Merazzi, M., Neglia, S., & Forzatti, P. (2005). Effect of ageing time on chemical and rheological evolution in γ-Al_2O_3 slurries for dip-coating. *Catalysis Today, 105,* 492–498.

Demirbas, A. (2002). Biodiesel from vegetable oils via transesterification in supercritical methanol. *Energy Conversion and Management, 43,* 2349–2356.

Demirbas, A. (2003). Biodiesel fuels from vegetable oils via catalytic and non-catalytic supercritical alcohol transesterifications and other methods: a survey. *Energy Conversion and Management, 44,* 2093–2109.

Demirbas, A. (2006). Biodiesel production via non-catalytic SCF method and biodiesel fuel characteristics. *Energy Conversion and Management, 47,* 2271–2282.

Demirbas, A. (2007). Biodiesel from sunflower oil in supercritical methanol with calcium oxide. *Energy Conversion and Management, 48,* 937–941.

Demirbas, A. (2008). Studies on cottonseed oil biodiesel prepared in noncatalytic SCF conditions. *Bioresource Technology, 99,* 1125–1130.

Demirbas, A. (2009). Biodiesel from waste cooking oil via base-catalytic and supercritical methanol transesterification. *Energy Conversion and Management, 50,* 923–927.

Deng, S. G., & Lin, Y. S. (1997). Granulation of sol-gel-derived nanostructured alumina. *AIChE Journal, 43,* 505–514.

Di, Y., Yu, Y., Sun, Y., Yang, X., Lin, S., Zhang, M., et al. (2003). Synthesis, characterization, and catalytic properties of stable mesoporous aluminosilicates assembled from preformed zeolite L precursors. *Microporous and Mesoporous Materials, 62,* 221–228.

Dıaz, I., Mohino, F., Pariente, J. P., & Sastre, E. (2003). Synthesis of MCM-41 materials functionalised with dialkylsilane groups and their catalytic activity in the esterification of glycerol with fatty acids. *Applied Catalysis A: General, 242,* 161–169.

Dıaz, I., Mohino, F., Perez-Pariente, J., & Satre, E. (2001). Synthesis, characterization and catalytic activity of MCM-41-type mesoporous silicas functionalized with sulfonic acid. *Applied Catalysis A: General, 205,* 19–30.

Dijk, V. H. J. M. P., & Schenk, W. J. (1984). Theoretical and experimental study of one-dimensional syneresis of a protein gel. *The Chemical Engineering Journal, 28,* 43–50.

Dmytryshyn, S. L., Dalai, A. K., Chaudhari, S. T., Mishra, H. K., & Reaney, M. J. (2004). Synthesis and characterization of vegetable oil derived esters: evaluation for their diesel additive properties. *Bioresource Technology, 92,* 55–64.

Donaldson, K., Li, X. Y., & MacNee, W. (1998). Ultrafine (nanometre) particle mediated lung injury. *Journal of Aerosol Science, 29,* 553–560.

Dossin, T. F., Reyniers, M. F., & Marin, G. B. (2006). Kinetics of heterogeneously MgO-catalyzed transesterification. *Applied Catalysis B: Environmental, 61,* 35–45.

Encinar, J. M., Gonzalez, J. F., & Rodriguez-Reinares, A. (2005). Biodiesel from used frying oil. Variables affecting the yields and characteristics of the biodiesel. *Industrial and Engineering Chemistry Research, 44,* 5491–5499.

Ensoz, S., Angın, D., & Yorgun, S. (2000). Influence of particle size on the pyrolysis of rapeseed (Brassica napus L.): fuel properties of bio-oil. *Biomass and Bioenergy, 19,* 271–279.

Fauchadour, D., Kolenda, F., Rouleau, L., Barre, L., & Normand, L. (2000). Peptization mechanisms of boehmite used as precursors for catalysts. *Studies in Surface Science and Catalysis, 143,* 453–461.

Feng, T., Gu, Z. B., & Jin, Z. Y. (2007). Chemical composition and some rheological properties of Mesona Blumes gum. *Food Science and Technology International, 13,* 55–61.

Feng, Y., He, B., Cao, Y., Li, J., Liu, M., & Yan, F. (2010). Biodiesel production using cation-exchange resin as heterogeneous catalyst. *Bioresource Technology, 101,* 1518–1521.

Fennema, O. R. (1975). *Freezing preservation. Principles of food science part II: Physical principles of food preservation* (pp. 173–215). Marcel Dekker: New York.

Ferretti, C. A., Olcese, R. N., Apesteguıa, C. R., & Di Cosimo, J. I. (2009). Heterogeneously Catalyzed Glycerolysis of Fatty Acid Methyl Esters: Reaction Parameter Optimization. *Industrial and Engineering Chemistry Research, 48,* 10387–10394.

Filip, V., Zajic, V., & Smidrkal, J. (1992). Methanolysis of rapeseed oil triglycerides. *Revue Francaise des Corps Gras, 39*, 91–94.

Freedman, B., Butterfield, R. O., & Pryde, E. H. (1986). Transesterification kinetics of soybean oil. *Journal of American Oil Chemistry Society, 63*, 1375–1380.

Freedman, B., & Pryde, E. H. (1984). Variables affecting the yields of fatty esters from transesterified vegetable oils. *Journal of the American Oil Chemists' Society, 61*, 1638–1643.

Freese, U., Heinrich, F., & Roessner, F. (1999). Acylation of aromatic compounds on H-Beta zeolites. *Catalysis Today, 49*, 237–244.

Furuta, S., Matsuhashi, H., & Arata, K. (2004). Biodiesel fuel production with solid superacid catalysis in fixed bed reactor under atmospheric pressure. *Catalysis Communications, 5*, 721–723.

Furuta, S., Matsuhashi, H., & Arata, K. (2006). Biodiesel fuel production with solid amorphous-zirconia catalysis in fixed bed reactor. *Biomass and Bioenergy, 30*, 870–873.

Garcia, C. M., Teixeira, S., Marciniuk, L. L., & Schuchardt, U. (2008). Transesterification of soybean oil catalyzed by sulfated zirconia. *Bioresource Technology, 99*, 6608–6613.

Gerpen, J. V. (2005). Biodiesel processing and production. *Fuel Processing Technology, 86*, 1097–1107.

Gerpen, V. J., & Knothe, G. (2005). Basics of the Transesterification Reaction. In G. Knothe, J. V. Gerpen, & J. Krahl (Eds.), *The Biodiesel Handbook* (pp. 26–41). Illinois: AOCS Press. Urbana.

Ghanem, A. (2003). The utility of cyclodextrins in lipase-catalyzed transesterification in organic solvents: enhanced reaction rate and enantioselectivity. *Organic and Biomolecular Chemistry, 1*, 1282–1291.

Goff, M. J., Bauer, N. S., Lopes, S., Sutterlin, W. R., & Suppes, G. J. (2004). Acid-catalyzed alcoholysis of soybean oil. *Journal of American Oil Chemistry Society, 81*, 415–420.

Granados, M. L., Alonso, D. M., Sadaba, I., & Ocon, P. (2009). Leaching and homogeneous contribution in liquid phase reaction catalysed by solids: The case of triglycerides methanolysis using CaO. *Applied Catalysis B: Environmental, 89*, 265–272.

Granados, M. L., Poves, M. D. Z., Alonso, D. M., Mariscal, R., Galisteo, F. C., Moreno-Tost, R., et al. (2007). Biodiesel from sunflower oil by using activated calcium oxide. *Applied Catalysis B: Environmental, 73*, 317–326.

Greenberg, B. (1989). Bragg's law with refraction. *Acta Cryst, 45*, 238–241.

Gregg, S. J., & Sing, K. S. (1982). *Adsorption. Surface Area and Porosity:* Academic Press, London.

Gui, M. M., Lee, K. T., & Bhatia, S. (2009). Supercritical ethanol technology for the production of biodiesel: Process optimization studies. *Journal of Supercritical Fluids, 49*, 286–292.

Gutierrez-Ortiz, J. I., Lopez-Fonseca, R., Gonzalez Ortiz de Elguea, C., Gonzalez- Marcos, M. P., & Gonzalez-Velasco, J. R. (2000). Mass transfer studies in the hydrogenation of methyl oleate over a Ni/SiO2 catalyst in the liquid phase. *Reaction Kinetics and Catalysis Letters, 70*, 341–348.

Haas, M. J., McAloon, A. J., Yee, W. C., & Foglia, T. A. (2006). A process model to estimate biodiesel production costs. *Bioresource Technology, 97*, 671–678.

Han, H., & Guan, Y. (2009). Synthesis of biodiesel from rapeseed oil using K_2O/γ-Al_2O_3 as nano-solid-base catalyst. *Journal of Natural Sciences, 14*, 75–79.

Hathaway, P. E., & Davis, M. E. (1989). Base catalysis by alkali-modified zeolites: I. Catalytic activity. *Journal of Catalysis, 116*, 263–278.

Hawash, S., Kamal, N., Zaher, F., Kenawi, O., & Diwani, G. E. (2009). Biodiesel fuel from Jatropha oil via non-catalytic supercritical methanol transesterification. *Fuel, 88*, 579–582.

He, C., Baoxiang, P., Dezheng, W., & Jinfu, W. (2007). Biodiesel production by the transesterification of cottonseed oil by solid acid catalysts. *Frontiers of Chemical Engineering in China, 1*, 11–15.

He, H., Sun, S., Wang, T., & Zhu, S. (2007a). Transesterification kinetics of soybean oil for production of biodiesel in supercritical methanol. *Journal of American Oil Chemistry Society, 84*, 399–404.

He, H., Wang, T., & Zhu, S. (2007b). Continuous production of biodiesel fuel from vegetable oi using supercritical methanol process. *Fuel, 86*, 442–447.

Hu, Z., & Srinivasan, M. P. (1999). Preparation of high-surface-area activated carbons from coconut shell. *Microporous and Mesoporous Materials, 27*, 11–18.

Ignat, M., Oers, C. J. V., Vernimmen, J., Mertens, M., Vermaak, S. P., Meynen, V., et al. (2010). Textural property tuning of ordered mesoporous carbon obtained by glycerol conversion using SBA-15 silica as template. *Carbon, 48*, 1609–1618.

Ilham, Z., & Saka, S. (2009). Dimethyl carbonate as potential reactant in non-catalytic biodiesel production by supercritical method. *Bioresource Technology, 100*, 1793–1796.

Jacobson, K., Gopinath, R., Meher, L. C., & Dalai, A. K. (2008). Solid acid catalyzed biodiesel production from waste cooking oil. *Applied Catalysis B: Environmental, 85*, 86–91.

Jegannathan, K. R., Abang, S., Poncelet, D., Chan, E. S., & Ravindra, P. (2008). Production of biodiesel using immobilized lipase-a critical review. *Critical Reviews in Biotechnology, 28*, 253–264.

Jiang, D. E., Zhao, B. Y., Xie, Y. C., Pan, G. C., Ran, G. P., & Min, E. Z. (2001). Structure and basicity of γ-Al_2O_3-supported MgO and its application to mercaptan oxidation. *Applied Catalysis A: General, 219*, 69–78.

Jitputti, J., Kitiyanan, B., Rangsunvigit, P., Bunyakiat, K., Attanatho, L., & Jenvanitpanjakul, P. (2006). Transesterification of crude palm kernel oil and crude coconut oil by different solid catalysts. *Chemical Engineering Journal, 116*, 61–66.

Johnson, M. F. L., & Mooi, J. (1968). The origin and type of pores in the alumina catalysts. *Journal of Catalysis, 10*, 342–354.

Jr., A. C. C., de Souza, L. K. C., da Costa, C. E. F., Longo, E., Zamian, J.R., da Rocha Filho, G. N. (2009). Production of biodiesel by esterification of palmitic acid over Mesoporous aluminosilicate Al-MCM-41. *Fuel, 88*, 461–468.

Kansedo, J., Lee, K. T., & Bhatia, S. (2009). Biodiesel production from palm oil via heterogeneous transesterification. *Biomass and Bioenergy, 33*, 271–276.

Kawashima, A., Matsubara, K., & Honda, K. (2008). Development of heterogeneous base catalysts for biodiesel production. *Bioresource technology, 99*(9), 34396–3443.

Kasim, N. S., Tsai, T. H., Gunawan, S., & Ju, Y. H. (2009). Biodiesel production from rice bran oil and supercritical methanol. *Bioresource Technology, 100*, 2399–2403.

Kawashim, A., Matsubara, K., & Honda, K. (2008). Development of heterogeneous base catalysts for biodiesel production. *Bioresource technology, 99*(9), 3439–3443.

Keyes, D. B. (1932). Esteri cation processes and equipment. *Industrial and Engineering Chemistry, 24*, 1096–1103.

Kim, S. S., Choi, J., & Kim, J. (2005). Plasma catalytic reaction of methane over nanostructured Ru/γ-Al2O3 catalysts in dielectric-barrier discharge. *Journal of Industrial and Engineering Chemistry, 11*, 533–539.

Kim, H. J., Kang, B. S., Kim, M. J., Park, Y. M., Kim, D. K., Lee, J. S., et al. (2004). Transesterification of vegetable oil to biodiesel using heterogeneous base catalyst. *Catalysis Today, 93–95*, 315–320.

King, C. J. (2007). *In Ullmann's Encyclopedia of Industrial Chemistry, Weinheim, electronic version.*

Kirkland, J. J. (1963). Fibrillar boehmite-A new adsorbent for gas solid chromatography. *Analytical Chemistry, 35*, 1295–1297.

Kirkland, J. J., Truszkowski, F. A., Dilks, C. H., Jr., & Engel, G. S. (2000). High pH mobile phase effects on silica-based reversed-phase high-performance liquid chromatographic columns. *Journal of Chromatography. A, 890*, 3–19.

Kiss, F. E., Jovanovic, M., & Boskovic, G. C. (2010). Economic and ecological aspects of biodiesel production over homogeneous and heterogeneous catalysts. *Fuel Processing Technology, 91*, 1316–1320.

Kiss, A. A., Omota, F., Dimian, A. C., & Rothenberg, G. (2006). The heterogeneous advantage: biodiesel by catalytic reactive distillation. *Topics in Catalysis, 40*, 141–150.

Knothe, G., Krahl, J., & Van Gerpen, J. (Eds.). (2005). *The biodiesel handbook*. Champaign, IL: AOCS Press.

Knothe, G., Sharp, C. A., & Ryan, T. W. (2006). Exhaust emissions of biodiesel, petrodiesel, neat methyl esters, and alkanes in a new technology engine. *Energy and Fuels, 20*, 403–408.

Knozinger, H. (1997). *Handbook of heterogeneous catalysis* (2nd ed., pp. 676–689). Weinheim: Wiley-VCH.

Kolaczkowski, S. T., Asli, U. A., & Davidson, M. G. (2009). A new heterogeneous ZnL2 catalyst on a structured support for biodiesel production. *Catalysis Today, 147*, 220–224.

Korytkowska, A., Barszczewska-Rybarek, I., & Gibas, M. (2001). Side-reactions in the transesterification of oligoethylene glycols by methacrylates. *Designed Monomers And Polymers, 4*, 27–37.

Kotwal, M. S., Niphadkar, P. S., Deshpande, S. S., Bokade, V. V., & Joshi, P. N. (2009). Transesterification of sunflower oil catalyzed by flyash-based solid catalysts. *Fuel, 88*, 1773–1778.

Kouzu, M., Kasuno, T., Tajik, M., Sugimoto, Y., Yamanaka, S., & Hidaka, J. (2008). Calcium oxide as a solid base catalyst for transesterification of soybean oil and its application to biodiesel production. *Fuel, 87*, 2798–2806.

Kresge, C. T., Leonowicz, M. E., Roth, W. J., Vartuli, J. C., & Beck, J. S. (1992). Ordered mesoporous molecular sieves synthesized by a liquid-crystal template mechanism. *Nature, 359*, 710–712.

Kusdiana, D., & Saka, S. (2001). Kinetics of transesterification in rapeseed oil to biodiesel fuel as treated in supercritical methanol. *Fuel, 80*, 693–698.

Kuznicki, S. M. (1989). *U. S. Patent No. 4,853,202*. Washington, DC: U.S. Patent and Trademark Office.

Kuznicki, S. M. (1991). Large-pored crystalline titanium molecular sieve zeolites, US Patent US Patent no. 5011591.

Lai, O. M., Ghazali, H. M., & Chong, C. L. (1999). Use of enzymatic transesterified palm stearin–sunflower oil blends in the preparation of table margarine formulation. *Food Chemistry, 64*, 83–88.

Lam, M. K., & Lee, K. T. (2010). Accelerating transesterification reaction with biodiesel as co-solvent: A case study for solid acid sulfated tin oxide catalyst. *Fuel, 89*, 3866–3870.

Lam, M. K., Tan, K. T., Lee, K. T., & Mohamed, A. R. (2009). Malaysian palm oil: surviving the food versus fuel dispute for a sustainable future. *Renewable and Sustainable Energy Reviews, 13*, 1456–1464.

Leclercq, E., Finiels, A., & Moreau, C. (2001). Transesterification of rapeseed oil in the presence of basic zeolites and related solid catalysts. *Journal of American Oil Chemistry Society, 78*, 1161–1165.

Lee, Y., Park, S. H., Lim, I. T., Han, K., & Lee, S. Y. (2000). Preparation of alkyl (R)-(2)-3-hydroxybutyrate by acidic alcoholysis of poly-(R)-(2)-3-hydroxybutyrate. *Enzyme and Microbial Technology, 27*, 33–36.

Lee, D.-W., Park, Y.-M., & Lee, K.-Y. (2009). Heterogeneous base catalysts for transesterification in biodiesel synthesis. *Catalysis Surveys from Asia, 13*, 63–77.

Lee, B. B., Ravindra, P., & Chan, E. S. (2009). New drop weight analysis for surface tension determination of liquids. *Colloids and Surfaces A: Physicochemical and Engineering Aspects, 332*, 112–120.

Leofanti, G., Padovan, M., Tozzola, G., & Venturelli, B. (1998). Surface area and pore texture of catalysts. *Catalysis Today, 41*, 207–219.

Leung, D. Y. C., & Guo, Y. (2006). Transesterification of neat and used frying oil: Optimization for biodiesel production. *Fuel Processing Technology, 87*, 883–890.

Levin, I., & Brandon, D. (1998). Metastable alumina polymorphs: Crystal structures and transition sequences. *Journal of the American Ceramic Society, 81*, 1995–2012.

Li, G., Smith, R. L., Jr., Inomata, H., & Arai, K. (2002). Synthesis and thermal decomposition of nitrate-free boehmite nanocrystals by supercritical hydrothermal conditions. *Materials Letters, 53*, 175–179.

Li, E., Xu, Z. P., & Rudolph, V. (2009). MgCoAl–LDH derived heterogeneous catalysts for the ethanol transesterification of canola oil to biodiesel. *Applied Catalysis B: Environmental, 88,* 42–49.

Lindlar, B., Luchinger, M., Haouas, M., Kogelbauer, A., Prins In, R., Galarneau, A., et al. (2001). Zeolites and mesoporous materials at the dawn of the 21st century. *Studies in Surface Science and Catalysis, 135,* 29–28.

Linko, Y. Y., Lamsa, M., Wu, X., Uosukainen, W., Sappala, J., & Linko, P. (1998). Biodegradable products by lipase biocatalysis. *Journal of Biotechnology, 66,* 41–50.

Linssen, T., Cassiers, K., Cool, P., & Vansant, E. F. (2003). Mesoporous templated silicates: An overview of their synthesis, catalytic activation and evaluation of the stability. *Advances in Colloid and Interface Science, 103,* 121–147.

Liu, K. S. (1994). Preparation of fatty-acid methyl esters for gas- chromatographic analysis of lipids in biological-materials. *Journal of American Oil Chemistry Society, 71,* 1179–1187.

Liu, X., He, H., Wang, Y., Zhu, S., & Piao, X. (2008b). Transesterification of soybean oil to biodiesel using CaO as a solid base catalyst. *Fuel, 87,* 216–221.

Liu, X., He, H., Wang, Y., & Zhu, S. (2007). Transesterification of soybean oil to biodiesel using SrO as a solid base catalyst. *Catalysis Communications, 8,* 1107–1111.

Liu, X., Piao, X., Wang, Y., Zhu, S., & He, H. (2008c). Calcium methoxide as a solid base catalyst for the transesterification of soybean oil to biodiesel with methanol. *Fuel, 87,* 1076–1082.

Liu, Q., Wang, A., Wang, X., Gao, P., Wang, X., & Zhang, T. (2008). Synthesis, characterization and catalytic applications of mesoporous γ-alumina from boehmite sol. *Microporous and Mesoporous Materials, 111,* 323–333.

Liu, R., Wang, X., Zhao, X., & Feng, P. (2008a). Sulfonated ordered mesoporous carbon for catalytic preparation of biodiesel. *Carbon, 46,* 1664–1669.

Liu, Y., Zhao, G., Liu, G., Wu, S., Chen, G., Zhang, W., et al. (2008d). Cyclopentadienyl-functionalized mesoporous MCM-41 catalysts for the transesterification of dimethyl oxalate with phenol. *Catalysis Communications, 9,* 2022–2025.

Lopez, D., Goodwin, J., Bruce, D., & Lotero, E. (2005). Transesterification of triacetin with methanol on solid acid and base catalysts. *Applied Catalysis A: General, 295,* 97–105.

Lopez, D. E., Suwannakarn, K., Bruce, D. A., & Goodwin, J. G., Jr. (2007). Esterification and transesterification on tungstated zirconia: Effect of calcination temperature. *Journal of Catalysis, 247,* 43–50.

Lotero, E., Goodwin, J. G., Bruce, D. A., Suwannakarn, K., Liu, Y., & Lopez, D. E. (2006). The catalysis of biodiesel synthesis. *Catalysis, 19,* 41–83.

Lotero, E., Liu, Y., Lopez, D. E., Suwannakarn, K., Bruce, D. A., & Goodwin, J. G., Jr. (2005). Synthesis of biodiesel via acid catalysis. *Industrial and Engineering Chemistry Research, 44,* 5353–5363.

Lou, W. Y., Zong, M. H., & Duan, Z. Q. (2008). Efficient production of biodiesel from high free fatty acid-containing waste oils using various carbohydrate-derived solid acid catalysts. *Bioresource Technology, 99,* 8752–8758.

Lukic, I., Krstic, J., Jovanovi, D., & Skala, D. (2009). Alumina/silica supported K2CO3 as a catalyst for biodiesel synthesis. *Bioresource Technology, 100,* 4690–4696.

Ma, F., & Hanna, M. A. (1999). Biodiesel production: A review. *Bioresource Technology, 70,* 1–15.

Ma, H., Li, S., Wang, B., Wang, R., & Tian, S. (2008). Transesterification of rapeseed oil for synthesizing biodiesel by K/KOH/γ-Al$_2$O$_3$ as heterogeneous base catalyst. *Journal of American Oil Chemistry Society, 85,* 263–270.

Mabaso, E. I., Van Steen, E., & Claeys, M. (2006). Fischer-Tropsch synthesis on supported iron crystallites of different size. *DGMK Tagungsbericht, 4,* 93–100.

Macario, A., Giordano, G., Onida, B., Cocina, D., Tagarelli, A., & Giuffre, A. M. (2010). Biodiesel production process by homogeneous/heterogeneous catalytic system using an acid–base catalyst. *Applied Catalysis A: General, 378,* 160–168.

Maceiras, R., Vega, M., Costa, C., Ramos, P., & Marquez, M. C. (2009). Effect of methanol content on enzymatic productionof biodiesel from waste frying oil. *Fuel, 88,* 2130–2134.

MacLeod, C. S., Harvey, A. P., Lee, A. F., & Wilson, K. (2008). Evaluation of the activity and stability of alkali-doped metal oxide catalysts for application to an intensified method of bio-diesel production. *Chemical Engineering Journal, 135*, 63–70.

Madras, G., Kolluru, C., & Kumar, R. (2004). Synthesis of biodiesel in supercritical fluids. *Fuel, 83*, 2029–2033.

Mani, T. V., Pillai, P. K., Damodaran, A. D., & Warrier, K. G. K. (1994). Dependence of calcination conditions of boehmite on the alumina particulate characteristics and sinterability. *Materials Letters, 19*, 237–241.

Martyanov, I. N., & Sayari, A. (2008). Comparative study of triglyceride transesterification in the presence of catalytic amounts of sodium, magnesium, and calcium methoxides. *Applied Catalysis A: General, 339*, 45–52.

Maskara, A., & Smith, D. M. (2005). Agglomeration during the drying of fine silica powders, Part II: The role of particle solubility. *Journal of the American Ceramic Society, 80*, 1715–1722.

Matatov, M. Y., & Sheintuch, M. (2002). Catalytic fibers and cloths. *Applied Catalysis A: General, 231*(1), 1–16.

Matsuda, H., & Okuhara, T. (1998). Catalytic synthesis of N-alkylacrylamide from acrylonitrile and 1-adamantanol with a novel solid heteropoly compound. *Catalysis Letters, 56*, 241–243.

Mazzocchia, C., Modica, G., Kaddouri, A., & Nannicini, R. (2004). Fatty acid methyl esters synthesis from triglycerides over heterogeneous catalysts in the presence of microwaves. *Comptes Rendus Chimie, 7*, 601–605.

Mbaraka, I. K., Radu, D. R., Lin, V. S. Y., & Shanks, B. H. (2003). Organosulfonic acid-functionalized mesoporous silicas for the esterification of fatty acid. *Journal of Catalysis, 219*, 329–336.

Mbaraka, I. K., & Shanks, B. H. (2006). Conversion of oils and fats using advanced mesoporous heterogeneous catalysts. *Journal of the American Oil Chemists' Society, 83*, 79–91.

McCarty, G. S., & Weiss, P. S. (1999). Scanning probe studies of single nanostructures. *Chemistry Review, 99*, 1983–1990.

McNeff, C. V., McNeff, L. C., Yan, B., Nowlan, D. T., Rasmussen, M., Gyberg, A. E., et al. (2008). A continuous catalytic system for biodiesel production. *Applied Catalysis A: General, 343*, 39–48

Meher, L. C., Kulkarni, M. G., Dalai, A. K., & Naik, S. N. (2006a). Transesterification of karanja (Pongamia pinnata) oil by solid basic catalysts. *European Journal of Lipid Science and Technology, 108*, 389–397.

Meher, L. C., Sagar, D. V., & Naik, S. N. (2006). Technical aspects of biodiesel production by transesterification: a review. *Renewable and Sustainable Energy Reviews, 10*, 248–268.

Meille, V. (2006). Review on methods to deposit catalysts on structured surfaces. *Applied Catalysis A: General, 315*, 1–17.

Mekhilef, S., Siga, S., & Saidur, R. (2011). A review on palm oil biodiesel as a source of renewable fuel. *Renewable and Sustainable Energy Reviews, 15*, 1937–1949.

Melde, B. J., Holland, B. T., Blandford, C. F., & Stein, A. (1999). Mesoporous sieves with unified hybrid inorganic/organic frameworks. *Chemistry of Materials, 11*, 3302–3308.

Melero, J. A., Grieken, R. V., & Morales, G. (2006). Advances in the synthesis and catalytic applications of organosulfonic-functionalized mesostructured materials. *Chemistry Review, 106*, 3790–3812.

Melero, J. A., Stucky, G. D., Van Grieken, R., & Morales, G. (2002). Direct syntheses of ordered SBA-15 mesoporous materials containing arenesulfonic acid groups. *Journal of Materials Chemistry, 12*, 1664–1670.

Meytal, Y. M., & Sheintuch, M. (2002). Catalytic fibers and cloths. *Applied Catalysis A: General, 231*, 1–16.

Mikolajczyk, T., Czapnik, D. W., & Bogun, M. (2004). Precursor alginate fibres containing nano-particles of SiO_2. *Fibres & Textiles in Eastern Europe, 12*, 19–23.

Mittelbach, M., & Remschmidt, C. (2004). *Biodiesels–the comprehensive handbook*. Graz, Austria: Karl-Franzens University.

Modica, C. M. G., Kaddouri, A., Nannicin, R. (2004). Fatty acid methyl esters synthesis from tri-glycerides over heterogeneous catalysts in the presence of microwaves. *C. R. Chimie 7,* 601–605.

Monica, C. G. A., lez, J. S. G., Robles, J. M. M., Tost, R. M., Castellon, E. R., Lopez, A. J., Azevedo, D. C. S., Jr., C. L. C., Torres, P. M. (2008). MgM (M = Al and Ca) oxides as basic catalysts in transesterification processes. *Applied Catalysis A: General. 347,* 162–168.

Mootabadi, H., Salamatinia, B., Bhatia, S., & Abdullah, A. Z. (2010). Ultrasonic-assisted biodiesel production process from palm oil using alkaline earth metal oxides as the heterogeneous cata-lysts. *Fuel, 89,* 1818–1825.

Morch, Y. A., Donati, I., Strand, B. L., & Skjak-Baek, G. (2006). Effect of Ca^{2+}, Ba^{2+}, and Sr^{2+} on alginate microbeads. *Biomacromolecules, 7,* 1471–1480.

Moreno, R., Salomoni, A., & Stamenkovic, I. (1997). Influence of slip rheology on pressure cast-ing of alumina. *Journal of the European Ceramic Society, 17,* 327–331.

MPOB (Malaysian Palm Oil Board). (2007). *Overview of the Malaysian oil palm industry.* Retrieved from www.mpob.gov.my.

Mul, G. & Moulijn, J. A. (2005). In J. A. Anderson & M. F. Garcia (Eds.), *Supported metals in catalysis* (pp. 1–32). London: Imperial College Press.

Nabeel, A., Jarrah, J. G., Ommen, V., & Lefferts, L. (2004). Immobilization of carbonnanofibers (CNFs). A new structured catalyst support. *Preprints of Papers- American Chemical Society, Division of Fuel Chemistry, 49,* 881–882.

Ngamcharussrivichai, C., Totarat, P., & Bunyakiat, K. (2008). Ca and Zn mixed oxide as a hetero-geneous base catalyst for transesterification of palm kernel oil. *Applied Catalysis A: General, 341,* 77–85.

Noiroj, K., Intarapong, P., Luengnaruemitchai, A., & Jai-In, S. (2009). A comparative study of KOH/Al_2O_3 and KOH/NaY catalysts for biodiesel production via transesterification from palm oil. *Renewable Energy, 34,* 1145–1150.

Noureddini, H., & Zhu, D. (1997). Kinetics of transesterification of soybean oil. *Journal of American Oil Chemistry Society, 74,* 1457–1463.

Pariente, J. P., Dıaz, I., Mohino, F., & Sastre, E. (2003). Selective synthesis of fatty monoglycer-ides by using functionalised mesoporous catalysts. *Applied Catalysis A: General, 254,* 173–188.

Passerini, S., Coustier, F., Giorgetti, M., & Smyrl, W. H. (1999). Li-Mn-O aerogels. *Electrochemical and Solid-State Letters, 2,* 483–485.

Peng, B., Shu, Q., Wang, J., Wang, G., Wang, D., & Han, M. (2008). Biodiesel production from waste oil feedstocks by solid acid catalysis. *Process Safety and Environmental Protection, 86,* 441–447.

Perego, G. (1998). Characterization of heterogeneous catalysts by X-ray diffraction techniques. *Catalysis Today, 41,* 251–259.

Perego, C., & Villa, P. (1997). Catalyst preparation methods. *Catalysis Today, 34,* 281–305.

Peyrin, F., Mastrogiacomo, M., Cancedda, R., & Martinetti, R. (2007). SEM and 3D synchrotron radiation micro-tomography in the study of bioceramic scaffolds for tissue-engineering appli-cations. *Biotechnology and Bioengineering, 97,* 638–648.

Philipse, A. P. (1993). Preparation of boehmite—silica colloids: Rods, spheres, needles and gels. *Colloids and Surfaces, A: Physicochemical and Engineering Aspects, 80,* 203–210.

Pierre, A. C., Elaloui, E., & Pajonk, G. M. (1998). Comparison of the structure and porous texture of alumina gels synthesized by different methods. *Langmuir, 14*(1), 66–73.

Pinna, F. (1998). Supported metal catalysts preparation. *Catalysis Today, 41,* 129–137.

Pinnarat, T., & Savage, P. E. (2008). Assessment of noncatalytic biodiesel synthesis using super-critical reaction conditions. *Industrial and Engineering Chemistry Research, 47,* 6801–6808.

Pinto, A. C., Guarieiro, L. N., Rezende, M. J., Ribeiro, N. M., Torres, E. A., Lopes, W. A., et al. (2005). Biodiesel: an overview. *Journal of the Brazilian Chemical Society, 16,* 1313–1330.

Popa, A. F., Rossignol, S., & Kappenstein, C. (2002). Ordered structure and preferred orientation of boehmite films prepared by the sol–gel method. *Journal of Non-Crystalline Solids, 306,* 169–174.

Portnoff, M. A., Purta, D. A., Nasta, M. A., Zhang, J. & Pourarian, F. (2006). *Methods for producing biodiesel*. PCT No. WO2006/002087.

Poulain, P. M., Warn-Varnas, A., & Niiler, P. P. (1996). Near-surface circulation of the Nordic seas as measured by Lagrangian drifters. *Journal of Geophysical Research: Oceans, 101*(C8), 18237–18258.

Prouzet, E., Khani, Z., Bertrand, M., Tokumoto, M., Guyot-Ferreol, V., & Tranchant, J. F. (2006). An example of integrative chemistry: Combined gelation of boehmite and sodium alginate for the formation of porous beads. *Microporous and Mesoporous Materials, 96*, 369–375.

Prouzet, E., Tokumoto, M. & Krivaya, A. (2004). Method for preparing beads containing a cross-linked mineral matrix. Patent No. WO2004009229

Qiu, F., Li, Y., Yang, D., Li, X. X., & Sun, P. (2011). Heterogeneous solid base nanocatalyst: Preparation, characterization and application in biodiesel production. *Bioresource Technology, 102*, 4150–4156.

Ramu, S., Lingaiah, N., Devi, B. L. A. P., Prasad, R. B. N., Suryanarayana, I., & Prasad, P. S. S. (2004). Esterification of palmitic acid with methanol over tungsten oxide supported on zirconia solid acid catalysts: Effect of method of preparation of the catalyst on its structural stability and reactivity. *Applied Catalysis A: General, 276*, 163–168.

Rao, M. A., & Anantheswaran, R. C. (1982). Rheology of fluids in food processing. *Journal of Food Technology, 36*, 116–126.

Rashtizadeh, E., Farzaneh, F., & Ghandi, M. (2010). A comparative study of KOH loaded on double aluminosilicate layers, microporous and mesoporous materials as catalyst for biodiesel production via transesterification of soybean oil. *Fuel, 89*, 3393–3398.

Ratanawilai, S. B., Suppalukpanya, K., & Tongurai, C. (2005). *Biodiesel from crude palm oil by sulfonated Vanadia-titania catalyst*. PSU-UNS International Conference on Engineering and Environment – ICEE.

Richardson, J. T. (1989). *Principles of catalyst development* (pp. 6–7). New York: Plenum Press.

Rouquerol, F., Rouquerol, J., & Sing, K. (1999). *Adsorption by powders and porous solids. Principles, methodology and application*. London: Academic Press. ISBN ISBN 0-12-598920- 2.

Royon, D., Daz, M., Ellenrieder, G., & Locatelli, S. (2007). Enzymatic production of biodiesel from cotton seed oil using t butanol as a solvent. *Bioresource Technology, 98*, 648–653.

Rueb, C. J., & Zukoski, C. F. (1992). Interparticle attractions and the mechanical properties of colloidal gels. *Materials Research Society Symposium Proceedings, 249*, 279–286.

Ryoo, R., Jun, S., Kim, J. M., & Jim, M. J. (1997). Generalised route to the preparation of mesoporous metallosilicates via post-synthetic metal implantation. *Chemical Communications, 22*, 2225–2226.

Saka, S., & Dadan, K. (2001). Biodiesel fuel from rapeseed oil as prepared in supercritical methanol. *Fuel, 80*, 225–231.

Samart, C., Sreetongkittikul, P., & Sookman, C. (2009). Heterogeneous catalysis of transesterification of soybean oil using KI/mesoporous silica. *Fuel Processing Technology, 90*, 922–925.

Sasidharan, M., & Kumar, R. (2004). Transesterification over various zeolites under liquid-phase conditions. *Journal of Molecular Catalysis A: Chemical, 210*, 93–98.

Schilling, C. H., Sikora, M., Tomasik, P., Li, C., & Garcia, V. (2002). Rheology of alumina nanoparticle suspensions: effects of lower saccharides and sugar alcohols. *Journal of the European Ceramic Society, 22*, 917–921.

Schuchardt, U., Serchelia, R., & Vargas, R. M. (1998). Transesterification of vegetable oils: A review. *Journal of the Brazilian Chemical Society, 9*, 199–210.

Schwab, A. W., Bagby, M. O., & Freedman, B. (1987). Preparation and properties of diesel fuels from vegetable oils. *Fuel, 66*, 1372–1378.

Schwarz, J. A., Contescu, C., & Contescu, A. (1995). Methods of preparation of catalytic materials. *Chemistry Review, 95*, 477–510.

Sercheli, R., Vargas, R. M., & Schuchardt, U. (1999). Alkyguanidine-catalyzed heterogeneous transesterification of soybean oil. *Journal of American Oil Chemistry Society, 76*, 1207–1210.

Serio, M. D., Cozzolino, M., Tesser, R., Patrono, P., Pinzari, F., Bonelli, B., et al. (2007). Vanadyl phosphate catalysts in biodiesel production. *Applied Catalysis A: General, 320*, 1–7.

Serio, M. D., Ledda, M., Cozzolino, M., Minutillo, G., Tesser, R., & Santacesaria, E. (2006). Transesterification of soybean oil to biodiesel by using heterogeneous basic catalysts. *Industrial and Engineering Chemistry Research, 45*, 3009–3014.

Serio, M. D., Tesser, R., Dimiccoli, M., Cammarota, F., Nastasi, M., & Santacesaria, E. (2005). Synthesis of biodiesel via homogeneous Lewis acid catalyst. *Journal of Molecular Catalysis A: Chemical, 239*, 111–115.

Serio, M. D., Tesser, R., Pengmei, L., & Santacesaria, E. (2008). Heterogeneous catalysts for biodiesel production. *Energy & Fuels, 22*, 207–217.

Serrano, A., Gallego, M., & Gonzalez, J. L. (2006). Assessment of natural attenuation of volatile aromatic hydrocarbons in agricultural soil contaminated with diesel fuel. *Environmental Pollution, 144*, 203–209.

Serwicka, E. M. (2000). Surface area and porosity, X-ray diffraction and chemical analyses. *Catalysis Today, 56*, 335–346.

Shah, S., & Gupta, M. N. (2007). Lipase catalyzed preparation of biodiesel from Jatropha oil in a solvent free system. *Process Biochemistry, 42*, 409–414.

Shah, P., Ramaswamy, A. V., Lazarc, K., & Ramaswamy, V. (2004). Synthesis and characterization of tin oxide-modified mesoporous SBA-15 molecular sieves and catalytic activity in transesterification reaction. *Applied Catalysis A: General, 273*, 239–248.

Sharma, L. D., Kumar, M., Saxena, A. K., Chand, M., & Gupta, J. K. (2002). Influence of pore size distribution on Pt dispersion in Pt-Sn/Al$_2$O$_3$ reforming catalyst. *Journal of Molecular Catalysis A: Chemical, 185*, 135–141.

Shu, Q., Zhang, Q., Xu, G., Nawaz, Z., Wang, D., & Wang, J. (2009). Synthesis of biodiesel from cottonseed oil and methanol using a carbon-based solid acid catalyst. *Fuel Processing Technology, 90*, 1002–1008.

Shumaker, J. L., Crofcheck, C., Tackett, S. A., Jimenez, E. S., Morgan, T., Ji, Y., et al. (2008). Biodiesel synthesis using calcined layered double hydroxide catalysts. *Applied Catalysis B: Environmental, 82*, 120–130.

Shuwen, L., Tong, C., Dongshen, T., Yi, Z., Yongcheng, L., & Gongyin, W. (2007). Synthesis of diphenyl carbonate via transesterification catalyzed by HMS mesoporous molecular sieves containing heteroelements. *Chinese Journal of Catalysis, 28*, 937–939.

Siakpas, P., Karagiannidis, A., & Theodoseli, M. (2006). *Biodiesel feedstock, production and uses* (World sustainable energy days). Austria: Wels.

Siladitya, B., Chatterjee, M., & Ganguli, D. (1999). Role of a surface active agent in the sol-emulsion-gel synthesis of spherical alumina powders. *Journal of Sol-Gel Science and Technology, 15*, 271–277.

Silva, C., Weschenfelder, T. A., Rovani, S., Corazza, F. C., Corazza, M. L., Dariva, C., et al. (2007). Continuous production of fatty acid ethyl esters from soybean oil in compressed ethanol. *Industrial and Engineering Chemistry Research, 46*, 5304–5309.

Singh, N. I., & Eipeson, W. E. (2000). Rheological behaviour of clarified mango juice concentrates. *Journal of Texture Studies, 31*, 287–295.

Singh, A. K. & Fernando, S. D. (2006a). *Catalyzed fasttransesterification of soybean oil using ultrasonication*. American Society of Agricultural Engineers, ASAE Annual Meeting, Portland, Oregon.

Singh, A. K., & Fernando, S. D. (2008). Transesterification of soybean oil using heterogeneous catalysts. *Energy & Fuels, 22*, 2067–2069.

Singh, A., He, B., Thompson, J., & Van Gerpen, J. (2006). Process optimization of biodiesel production using different alkaline catalysts. *Applied Engineering in Agriculture, 22*, 597–600.

Smith, G. V., & Notheisz, F. (2006). *Heterogeneous catalysis in organic chemistry*. New York, NY: Academic Press Inc.

Song, K. C., & Chung, I. J. (1989). Rheological properties of aluminium hydroxide sols during sol-gel transition. *Journal of Non-Crystalline Solids, 107*, 193–198.

Sreeprasanth, P. S., Srivastava, R., Srinivas, D., & Ratnasamy, P. (2006). Hydrophobic, solid acid catalysts for production of biofuels and lubricants. *Applied Catalysis A: General, 314*, 148–159.

Stamenkovic, O. S., Lazic, M. L., Todorovic, Z. B., Veljkovic, V. B., & Skala, D. U. (2007). The effect of agitation intensity on alkali-catalyzed methanolysis of sunflower oil. *Bioresource Technology, 98*, 2688–2699.

Stern, R. & Hillion, G. (1990). Purification of esters. European Patent Application No. EP 356317.

Storck, S., Bretinger, H., & Maier, W. F. (1998). Characterization of micro and mesoporous solids by physisorption methods and pore-size analysis. *Applied Catalysis, A: General, 174*, 137–146.

Sun, L. B., Chun, Y., Gu, F. N., Yue, M. B., Yu, Q., & Wang, Y. (2007). A new strategy to generate strong basic sites on neutral salt KNO3 modified NaY. *Materials Letters, 61*, 2130–2134.

Sun, L. B., Gong, L., Liu, X. Q., Gu, F. N., Chun, Y., & Zhu, J. H. (2009). Generating basic sites on zeolite Y by potassium species modification: Effect of base precursor. *Catalysis Letters, 132*, 218–224.

Suppes, G. J., Dasari, M. A., Doskocil, E. J., Mankidy, P. J., & Goff, M. J. (2004). Transesterification of soybean oil with zeolite and metal catalysts. *Applied Catalysis A: General, 257*, 213–223.

Tamalampudi, S., Talukder, M. R., Hamad, S., Numata, T., Kondo, A., & Fukuda, H. (2008). Enzymatic production of biodiesel from Jatropha oil: A comparative study of immobilized-whole cell and commercial lipases as a biocatalyst. *Biochemical Engineering Journal, 39*, 185–189.

Tanabe, K., Misono, M., Ono, Y., & Hattori, H. (1989). *New solid acids and bases*. Amsterdam: Elsevier.

Tateno, T. & Sasaki, T. (2004). Process for producing fatty acid fuels comprising fatty acids esters. U.S. Patent 6818026.

Tantirungrotechai, J., Chotmongkolsap, P., & Pohmakotr, M. (2010). Synthesis, characterization, and activity in transesterification of mesoporous Mg–Al mixed metal oxides. *Microporous and Mesoporous Materials, 128*(1), 41–47.

Taylor, G. I. (1966). Studies in electrohydrodynamics 1. The circulation produced in a drop by an electric field. *Proceedings of the Royal Society of London, 291*, 159–166.

Tsai, M. S., & Yung, F. H. (2007). Boehmite modification of nano grade α-alumina and the rheological properties of the modified slurry. *Ceramics International, 33*, 739–745.

Vazquez, A., Lopez, T., Gomez, R., Bokhimit, M. A., & Novarot, O. (1997). X-ray diffraction, FTIR, and NMR characterization of sol-gel alumina doped with lanthanum and cerium. *Journal of Solid State Chemistry, 128*, 161–168.

Verziu, M., Florea, M., Simon, S., Simon, V., Filip, P., Parvulescu, V. I., et al. (2009). Transesterification of vegetable oils on basic large mesoporous alumina supported alkaline fluorides—Evidences of the nature of the active site and catalytic performances. *Journal of Catalysis, 263*, 56–66.

Vicente, G., Coteron, A., Martinez, M., & Aracil, J. (1998). Application of the factorial design of experiments and response surface methodology to optimize biodiesel production. *Industrial Crops and Products, 8*, 29–35.

Vyas, A. P., Subrahmanyam, N., & Patal, P. A. (2009). Production of biodiesel through transesterification of Jatropha oil using KNO_3/Al_2O_3 solid catalyst. *Fuel, 88*, 625–628.

Wahab, M. A., & Ha, C. S. (2005). Ruthenium-functionalised hybrid periodic mesoporous organosilicas: Synthesis and structural characterization. *Journal of Materials Chemistry A, 15*, 508–516.

Wan, T., Yu, P., Gong, S., Li, Q., & Luo, Y. (2008). Application of KF/MgO as a heterogeneous catalyst in the production of biodiesel from rapeseed oil. *Korean Journal of Chemical Engineering, 25*, 998–1003.

Wan, T., Yu, P., Wang, S., & Luo, Y. (2009). Application of sodium aluminate as a heterogeneous base catalyst for biodiesel production from soybean oil. *Energy & Fuels, 23*, 1089–1092.

Wang, Z. M., & Lin, Y. S. (1998). Sol-gel synthesis of pure and copper oxide coated mesoporous alumina granular particles. *Journal of Catalysis, 174*, 43–51.

Wang, Y., Ou, S., Liu, P., Xue, F., & Tang, S. (2006). Comparison of two different processes to synthesize biodiesel by waste cooking oil. *Journal of Molecular Catalysis A: Chemical, 252,* 107–112.

Wang, L., & Yang, J. (2007). Transesterification of soybean oil with nano-MgO or not in supercritical and subcritical methanol. *Fuel, 86,* 328–333.

Wang, Y., Zhang, F., Xu, S., Yang, L., Li, D., Evans, D. G., et al. (2008). Preparation of macrospherical magnesia-rich magnesium aluminate spinel catalysts for methanolysis of soybean oil. *Chemical Engineering Science, 63,* 4306–4312.

Ward, D. A., & Ko, E. L. (1995). Sol-gel synthesis of zirconia supports. Important properties for generating n-butane isomerization activity upon sulfate promotion. *Journal of Catalysis, 157,* 321–333.

Wei, Z., Xu, C., & Li, B. (2009). Application of waste eggshell as low-cost solid catalyst for biodiesel production. *Bioresource Technology, 100,* 2883–2885.

Wen, L., Wang, Y., Lu, D., Hu, S., & Han, H. (2010). Preparation of KF/CaO nanocatalyst and its application in biodiesel production from Chinese tallow seed oil. *Fuel, 89,* 2267–2271.

Williams, J. L. (2001). Monolith structures, materials, properties and uses. *Catalysis Today, 69,* 3–9.

Xiao, F. S. (2004). Hydrothermally stable and catalytically active ordered mesoporous materials assembled from preformed zeolites nanoclusters. *Catalysis Surveys from Asia, 35,* 151–159.

Xie, W., & Huang, X. (2006). Synthesis of biodiesel from soybean oil using heterogeneous kf/zno catalyst. *Catalysis Letters, 107,* 53–59.

Xie, W., & Li, H. (2006a). Alumina-supported potassium iodide as a heterogeneous catalyst for biodiesel production from soybean oil. *Journal of Molecular Catalysis A: Chemical, 255,* 1–9.

Xie, W. L., Peng, H., & Chen, L. G. (2006a). Transesterification of soybean oil catalyzed by potassium loaded on alumina as a solid-base catalyst. *Applied Catalysis A: General, 300,* 67–74.

Xie, W., Peng, H., & Chen, L. (2006b). Calcined Mg–Al hydrotalcites as solid base catalysts for methanolysis of soybean oil. *Journal of Molecular Catalysis A: Chemical, 246,* 24–32.

Xie, W., Yang, Z., & Chun, H. (2007). Catalytic properties of lithium-doped ZnO catalysts used for biodiesel preparations. *Industrial and Engineering Chemistry Research, 46,* 7942–7949.

Xin, B. H., Zhen, S. X., Hua, L. X., & Yong, L. S. (2009). Synthesis of porous CaO microsphere and its application in catalyzing transesterification reaction for biodiesel. *Transactions of the Nonferrous Metals Society of China, 19,* 674–677.

Xu, L., Li, W., Hu, J., Yang, X., & Guo, Y. (2009). Biodiesel production from soybean oil catalyzed by multifunctionalized $Ta_2O_5/SiO_2-[H_3PW_{12}O_{40}/R]$ (R = Me or Ph) hybrid catalyst. *Applied Catalysis B: Environmental, 90,* 587–594.

Yamaguchi, T. (1990). Recent progress in solid superacid. *Applied Catalysis A: General, 61,* 1–25.

Yan, S., Lu, H., & Liang, B. (2008). Supported CaO catalysts used in the transesterification of rapeseed oil for the purpose of biodiesel production. *Energy & Fuels, 22,* 646–651.

Yan, S., Salley, S. O., & Ng, K. Y. S. (2009). Simultaneous transesterification and esterification of unrefined or waste oils over $ZnO-La_2O_3$ catalysts. *Applied Catalysis A: General, 353,* 203–212.

Yang, Z., & Lin, Y. S. (2000). Sol-Gel synthesis of silica/γ-alumina granules. *Industrial and Engineering Chemistry Research, 39,* 4944–4948.

Yap, Y. H. T., Lee, H. V., Hussein, M. Z., & Yunus, R. (2011). Calcium-based mixed oxide catalysts for methanolysis of Jatropha curcas oil to biodiesel. *Biomass and Bioenergy, 35,* 827–834.

Yin, J. Z., Xiao, M., & Song, J. B. (2008). Biodiesel from soybean oil in supercritical methanol with co-solvent. *Energy Conversion and Management, 49,* 908–912.

Yin, S. F., Zhang, Q. H., Xu, B. Q., Zhu, W. X., Ng, C. F., & Au, C. T. (2004). Investigation on the catalysis of COx-free hydrogen generation from ammonia. *Journal of Catalysis, 224,* 384–396.

Yoldas, B. E. (1975). Alumina sol preparation from alkoxides. *American Ceramic Society Bulletin, 51,* 289–290.

Yoosuk, B., Udomsap, P., Puttasawat, B., & Krasae, P. (2010). Modification of calcite by hydration–dehydration method for heterogeneous biodiesel production process: The effects of water on properties and activity. *Chemical Engineering Journal, 162*, 135–141.

Zabeti, M., Daud, W. M. A. W., & Aroua, M. K. (2010). Biodiesel production using alumina-supported calcium oxide: an optimization study. *Fuel Processing Technology, 91*, 243–248.

Zhang, Y., Dube, M. A., McLean, D. D., & Kates, M. (2003). Biodiesel production from waste cooking oil: 1. Process design and technological assessment. *Bioresource Technology, 89*, 1–16.

Zheng, S., Kates, M., Bube, M. A., & McLean, D. D. (2006). Acid-catalyzed production of biodiesel from waste frying oil. *Biomass and Bioenergy, 30*, 267–272.

Zheng, Y., Wua, X. M., Christopher, B. W., Jing, Q., & Zhu, L. M. (2009). Dual response surface-optimized process for feruloylated diacylglycerols by selective lipase-catalyzed transesterification in solvent free system. *Bioresource Technology, 100*, 2896–2901.

Zong, M. H., Duan, Z. Q., Lou, W. Y., Smith, T. J., & Wu, H. (2007). Preparation of a sugar catalyst and its use for highly efficient production of biodiesel. *Green Chemistry, 9*, 434–437.

Printed in the United States
By Bookmasters